AF883141

Methods in Molecular Biology

Series Editor
John M. Walker
School of Life and Medical Sciences
University of Hertfordshire
Hatfield, Hertfordshire, UK

For further volumes:
http://www.springer.com/series/7651

For over 35 years, biological scientists have come to rely on the research protocols and methodologies in the critically acclaimed *Methods in Molecular Biology* series. The series was the first to introduce the step-by-step protocols approach that has become the standard in all biomedical protocol publishing. Each protocol is provided in readily-reproducible step-by-step fashion, opening with an introductory overview, a list of the materials and reagents needed to complete the experiment, and followed by a detailed procedure that is supported with a helpful notes section offering tips and tricks of the trade as well as troubleshooting advice. These hallmark features were introduced by series editor Dr. John Walker and constitute the key ingredient in each and every volume of the *Methods in Molecular Biology* series. Tested and trusted, comprehensive and reliable, all protocols from the series are indexed in PubMed.

Plant Stem Cells

Methods and Protocols

Edited by

Muhammad Naseem

Department of Life and Environmental Sciences, College of Natural and Health Sciences, Zayed University, Abu Dhabi, UAE; Functional Genomics and Systems Biology Group, Department of Bioinformatics, Biocenter, University of Würzburg, Würzburg, Germany

Thomas Dandekar

Department of Bioinformatics, University of Würzburg, Würzburg, Germany

Humana Press

Editors
Muhammad Naseem
Department of Life
and Environmental Sciences
College of Natural and Health
Sciences, Zayed University
Abu Dhabi, UAE

Functional Genomics
and Systems Biology Group
Department of Bioinformatics
Biocenter, University of Würzburg
Würzburg, Germany

Thomas Dandekar
Department of Bioinformatics
University of Würzburg
Würzburg, Germany

ISSN 1064-3745 ISSN 1940-6029 (electronic)
Methods in Molecular Biology
ISBN 978-1-0716-0182-2 ISBN 978-1-0716-0183-9 (eBook)
https://doi.org/10.1007/978-1-0716-0183-9

© Springer Science+Business Media, LLC, part of Springer Nature 2020
This work is subject to copyright. All rights are reserved by the Publisher, whether the whole or part of the material is concerned, specifically the rights of translation, reprinting, reuse of illustrations, recitation, broadcasting, reproduction on microfilms or in any other physical way, and transmission or information storage and retrieval, electronic adaptation, computer software, or by similar or dissimilar methodology now known or hereafter developed.
The use of general descriptive names, registered names, trademarks, service marks, etc. in this publication does not imply, even in the absence of a specific statement, that such names are exempt from the relevant protective laws and regulations and therefore free for general use.
The publisher, the authors, and the editors are safe to assume that the advice and information in this book are believed to be true and accurate at the date of publication. Neither the publisher nor the authors or the editors give a warranty, express or implied, with respect to the material contained herein or for any errors or omissions that may have been made. The publisher remains neutral with regard to jurisdictional claims in published maps and institutional affiliations.

This Humana imprint is published by the registered company Springer Science+Business Media, LLC, part of Springer Nature.
The registered company address is: 233 Spring Street, New York, NY 10013, U.S.A.

Preface

Owing to their sessile nature, plants are devoid of mechanical articulation in parts of their bodies, and hence they are comparatively more vulnerable to biotic and abiotic environmental stresses. By default, they also are programmed to replace lost body organs such as leaves and flowers on a regular basis throughout their life. The pluripotent stem cells in plants constantly provide precursor cells to form differentiated tissues and body organs. In plants, the shoot apical meristem (SAM), the root apical meristem (RAM), and the vascular meristem are the custodians of stem cells. These stem cell niches maintain a specific signaling environment to stop them from entering differentiation all at once. However, a required number of undifferentiated stem cells are also kept through a process of self-renewal. Being a custodian of the next generation of plants through seeds and flowers, the stem cell niches constantly supply cells to meet the programming and contingency requirements. Due to constant improvement technology and with the advent of high throughput methods, we are gaining deeper insights into mechanisms that control the aspects of plant stem cell signaling events in their respective niches. These conceptual developments are meticulously covered in the mainstream journals of biological sciences. However, methodological approaches in terms of bench protocols and associated experimental pitfalls are always described in little details in high-ranking journal owing to space constraints. To address this issue and provide researchers with real-time laboratory protocols with more personal insight from various groups actively working on plant stem cells research, we aim to collect these important methods. This collection of vital protocols will be of a practical importance to address the needs of students, technicians as well as advanced researchers addressing the research theme of stem cells in plant biology.

Abu Dhabi, UAE *Muhammad Naseem*
Würzburg, Germany *Thomas Dandekar*

Contents

Contributors

FATIMA A. ALREMEITHI • *Department of Life and Environmental Sciences, College of Natural and Health Sciences, Zayed University, Abu Dhabi, UAE*

GÉRALDINE BRUNOUD • *Laboratoire Reproduction et Développement des Plantes (RDP), Univ Lyon, ENS de Lyon, UCB Lyon 1, CNRS, INRA, Lyon, France*

ONUR CELIK • *Molecular Biology and Genetics Department, Faculty of Science, Mugla Sitki Kocman University, Mugla, Turkey*

THOMAS DANDEKAR • *Department of Bioinformatics, Biocenter, Functional Genomics and Systems Biology Group, University of Würzburg, Würzburg, Germany*

JOACHIM FORNER • *Organelle Biology, Max Planck Institute of Molecular Plant Physiology, Potsdam, Germany*

RIEKO FUJINAMI • *Faculty of Education, Kyoto University of Education, Kyoto, Japan*

SELIN GALATALI • *Molecular Biology and Genetics Department, Faculty of Science, Mugla Sitki Kocman University, Mugla, Turkey*

CARLOS S. GALVAN-AMPUDIA • *Laboratoire Reproduction et Développement des Plantes (RDP), Univ Lyon, ENS de Lyon, UCB Lyon 1, CNRS, INRA, Lyon, France*

SEVINC GULDAG • *Molecular Biology and Genetics Department, Faculty of Science, Mugla Sitki Kocman University, Mugla, Turkey*

GÖKHAN GUN • *Department of Molecular Biology and Genetics, Bogazici University, Istanbul, Turkey*

HUSNA • *Department of Botany, Garden Campus, Abdul Wali Khan University, Mardan, Khyber Pakhtunkhwa, Pakistan*

FARES M. HOWARI • *Department of Life and Environmental Sciences, College of Natural and Health Sciences, Zayed University, Abu Dhabi, UAE*

ANWAR HUSSAIN • *Department of Botany, Garden Campus, Abdul Wali Khan University, Mardan, Khyber Pakhtunkhwa, Pakistan*

JIBRAN IQBAL • *Department of Life and Environmental Sciences, College of Natural and Health Sciences, Zayed University, Abu Dhabi, UAE*

BADSHAH ISLAM • *Department of Agriculture, Garden Campus, Abdul Wali Khan University Mardan, Mardan, Khyber Pakhtunkhwa, Pakistan*

YULING JIAO • *State Key Laboratory of Plant Genomics and National Center for Plant Gene Research (Beijing), Institute of Genetics and Developmental Biology, The Innovative Academy of Seed Design, Chinese Academy of Sciences, Beijing, China; National Center for Plant Gene Research, Beijing, China; University of Chinese Academy of Sciences, Beijing, China*

MARTIN KALTDORF • *Department of Molecular Biology and Genetics, University of Wuerzburg, Wuerzberg, Germany*

ERGUN KAYA • *Molecular Biology and Genetics Department, Mugla Sitki Kocman University, Mugla, Turkey*

VIKTORIYA V. LAVREKHA • *Institute of Cytology and Genetics SB RAS, Novosibirsk, Russia; LCTEB, Novosibirsk State University, Novosibirsk, Russia*

Wu Liu • *National Key Laboratory of Plant Molecular Genetics, CAS Center for Excellence in Molecular Plant Sciences, Institute of Plant Physiology and Ecology, Shanghai Institutes for Biological Sciences, Chinese Academy of Sciences, Shanghai, China*

Monika Mahajan • *Indian Institute of Science Education and Research Mohali, SAS Nagar, India*

Victoria V. Mironova • *Institute of Cytology and Genetics SB RAS, Novosibirsk, Russia; LCTEB, Novosibirsk State University, Novosibirsk, Russia*

Waheed Murad • *Department of Botany, Garden Campus, Abdul Wali Khan University Mardan, Mardan, Khyber Pakhtunkhwa, Pakistan*

Muhammad Naseem • *Department of Life and Environmental Sciences, College of Natural and Health Sciences, Zayed University, Abu Dhabi, UAE; Functional Genomics and Systems Biology Group, Department of Bioinformatics, Biocenter, University of Würzburg, Würzburg, Germany*

Ozge Osmanoglu • *Department of Bioinformatics, Biocenter, University of Würzburg, Würzburg, Germany*

Klaus Palme • *Institute of Biology II/Molecular Plant Physiology, Centre for BioSystems Analysis, BIOSS Centre for Biological Signalling Studies University of Freiburg, Freiburg, Germany*

Frantisek Paprstein • *Department of Fruit Genebanks, Research and Breeding Institute of Pomology Holovousy Ltd., Horice, Czech Republic*

Taras Pasternak • *Institute of Biology II/Molecular Plant Physiology, Centre for BioSystems Analysis, BIOSS Centre for Biological Signalling Studies University of Freiburg, Freiburg, Germany*

Aleksandra Sapala • *Biosystems Science and Engineering, ETH Zurich, Mattenstrasse, Basel, Switzerland*

Jiri Sedlak • *Department of Fruit Genebanks, Research and Breeding Institute of Pomology Holovousy Ltd., Horice, Czech Republic*

Bihai Shi • *State Key Laboratory of Plant Genomics and National, Center for Plant Gene Research (Beijing), Institute of Genetics and Developmental Biology, The Innovative Academy of Seed Design, Chinese Academy of Sciences, Beijing, China; National Center for Plant Gene Research, Beijing, China; University of Chinese Academy of Sciences, Beijing, China*

Richard S. Smith • *John Innes Centre, Computational and Systems Biology, Norwich Research Park, Norwich, UK*

Mugdha Srivastava • *Department of Bioinformatics, Biocenter, University of Wuerzburg, Wuerzburg, Germany*

Aqib Syed • *Key Laboratory of Plant Cell Engineering and Germplasm Innovation, Ministry of Education, School of Life Sciences, Shandong University, Jinan, China*

Ihsan Ullah • *Department of Environmental Science, Islamic International University Islamabad, Islamabad, Pakistan*

Teva Vernoux • *Laboratoire Reproduction et Développement des Plantes (RDP), Univ Lyon, ENS de Lyon, UCB Lyon 1, CNRS, INRA, Lyon, France*

Hongli Wang • *State Key Laboratory of Plant Genomics and National Center for Plant Gene Research (Beijing), Institute of Genetics and Developmental Biology, The Innovative*

Academy of Seed Design, Chinese Academy of Sciences, Beijing, China; University of Chinese Academy of Sciences, Beijing, China

LIN XU • *National Key Laboratory of Plant Molecular Genetics, CAS Center for Excellence in Molecular Plant Sciences, Institute of Plant Physiology and Ecology, Shanghai Institutes for Biological Sciences, Chinese Academy of Sciences, Shanghai, China; University of Chinese Academy of Sciences, Beijing, China*

RAM KISHOR YADAV • *Indian Institute of Science Education and Research Mohali, SAS Nagar, India*

JIE YU • *National Key Laboratory of Plant Molecular Genetics, CAS Center for Excellence in Molecular Plant Sciences, Institute of Plant Physiology and Ecology, Shanghai Institutes for Biological Sciences, Chinese Academy of Sciences, Shanghai, China*

NING ZHAI • *National Key Laboratory of Plant Molecular Genetics, CAS Center for Excellence in Molecular Plant Sciences, Institute of Plant Physiology and Ecology, Shanghai Institutes for Biological Sciences, Chinese Academy of Sciences, Shanghai, China; University of Chinese Academy of Sciences, Beijing, China*

Chapter 1

Regulatory Role of Phytohormones in Maintaining Stem Cells and Boundaries of Stem Cell Niches

Aqib Syed, Anwar Hussain, Waheed Murad, and Badshah Islam

Abstract

Plants are multicellular organism composed of different types of cells. These all kinds of cells are formed from pluripotent stem cells present at different positions in plant called stem cell niches. All these stem cell niches and their boundaries are maintained by complex regulatory mechanism at molecular level involving different genes, cofactors, and phytohormones. In this chapter, we discussed the regulatory mechanism and models of stem cell maintenance, specifying their boundaries at different stem cell niches.

Key words Stem cell niches, *Arabidopsis*, Cytokinin, Auxin and jasmonic acid

1 Shoot Stem Cell Niches

Multicellular mode of life begins with a single cell which develops into multicellular body via multiple rounds of cell divisions. Plants are among the multicellular organisms living thousands of years. During their long journey of life, they pass through repeated cycles of developmental stages and witness endless mechanical injuries which contribute to their amazing power of regeneration and recovery of dead cells by producing new cells. This ability is due to the stem cells which remain undifferentiated and undergo several rounds of divisions. Locations of these stem cells within the plants are called stem cell niches. In plants, the stem cell niches are called meristems which foster the survival of stem cells and also the production of the progeny cells fated for differentiation. The meristem mostly lies at the tips of the plants, including SAM and RAM. Two additional meristems called vascular cambium and cork cambium are the lateral meristems of plant body, which are involved in secondary growth. The ring-shaped vascular cambium produces xylem and phloem, while cork cambium's responsibility is the replenishment of the regularly shedding outer layer called bark. Besides other functions, all kinds of meristems have two basic

Muhammad Naseem and Thomas Dandekar (eds.), *Plant Stem Cells: Methods and Protocols*, Methods in Molecular Biology, vol. 2094, https://doi.org/10.1007/978-1-0716-0183-9_1, © Springer Science+Business Media, LLC, part of Springer Nature 2020

functions: the production of new cells and the initiation of organ formation [1]. The mechanism of the production of new cells has been demonstrated by Satina and coworkers in 1940 by cell-tracking experiments in the shoot meristem. Through the treatment of colchicine, they produced polyploidy in the single cell of *Datura* and observed that the shoot meristem is comprise of clonally separated three layers of cells named as L1 (epidermal layer), L2 (subepidermal layer), and L3 (inner layer) [2]. Three layers of the shoot meristem is typical characteristic of dicots, but this varies in other groups of plants, e.g., in monocot, it consists of two layers; gymnosperm, only one layer; and, in lower plants (bryophytes and ferns), these layers are absent and all of their cells are from single apical cell. Later on, in 1970, Stewart and Dermen found that one-third of each layer is originated from one single cell which indicates the presences of three stem cells in one layer [3]. Stem cells can produce new cells and remain undifferentiated when they remain in a location having specific environment. Interestingly, the fate of the daughter stem cells does not depend on their parental cell but on the environment where they locate, and their displacement among different zones modulates their behavior according to their new destination, e.g., when L1 cells are displaced to the periphery of L2, they change to L2-type cells. This interesting phenomenon suggests the existence of some kind of chemical signals that maintain the ability of division, and cells moving away from the influence of these signals are bound to differentiation [1].

Plants survive for many years with continuous growth and replace old and dead cells by cell division throughout their life span. Here, question arrived that how the plant stem cell deals with mutation created during replication and survives for hundreds of years without accumulating mutation. Possible answer of this question is the infrequent division of the stem cell and the finite number of division of the stem cell daughter cells before their displacement. These two reasons reduce the chance of mutation produced during DNA replication [3, 4]. Besides that, DNA present in the root and shoot meristem cells is supersensitive to DNA damage resulting in death of the cell's damaged DNA, keeping stem cell system clean from the compromised DNA [5].

1.1 Structure of Shoot Apical Meristem

The shoot apical meristem (SAM) of *Arabidopsis* is differentiated into three distinct zones including the peripheral zone (PZ), central zone (CZ), and rib zone (RZ). Stem cells of the peripheral zones undergo rapid divisions as compared to that of the central zone. These rapidly dividing stem cells found in the peripheral zone are responsible for and are source of the lateral organs in mature shoot. In the rib zone, the cells start to assume a flattened shape to initiate the differentiation toward the central stem tissue (Fig. 1). The individual divisions of the shoot meristem stem cells produce two cells among which one acts as the stem cell, while the other

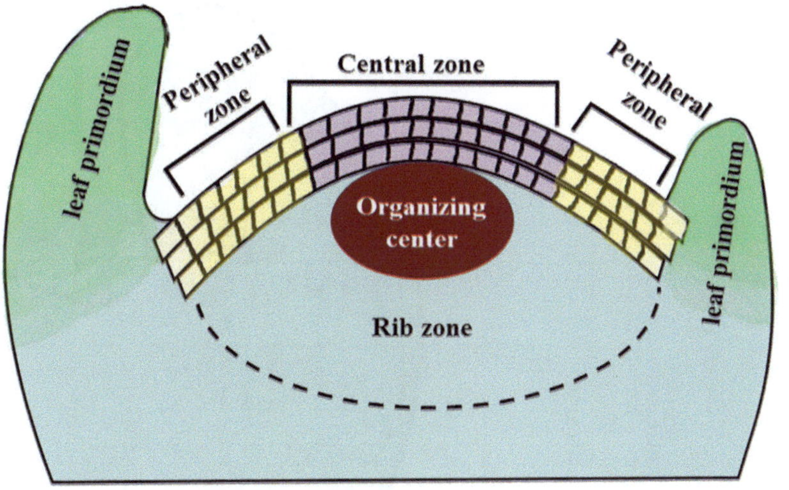

Fig. 1 Graphical presentation of the shoot apical meristem showing different zones

undergoes differentiation depending on their positions [6, 7]. How stem cells maintain their identity and how the boundaries of stem cell niches are maintained in the meristem zones are difficult questions to answer. Investigation of the past decade provides adequate knowledge to understand the basic molecular mechanism of stem cell regulation.

1.2 STM Maintains the Shoot Meristem by Regulating Cytokinin and Gibberellic Acid

The first gene (*KN*) known to be involved in plant stem cell regulation was identified in maize [8]. This gene is a member of *KNOTTED1-like homeobox* (KNOX) gene family. Normally, *KN* gene is expressed in undifferentiated cell in meristematic region but absent in leaf anlagen cells indicating its key role in maintaining the stem cells undifferentiated [9]. In *Arabidopsis*, *SHOOTMERISTEMLESS (STM)* shows similar pattern of expression, and mutation in this gene leads to the absence of the shoot meristem [10] and fused cotyledons. Mutant of the *STM* also failed to initiate the meristem at postembryonic stage. Study of *STM* mutant suggests that *STM* gene has two functions; the first is to prevent the cell differentiation in the meristematic region, and the second is the repression of cell division between the lateral organs to keep them separated [10–12]. STM gene regulates cytokinin (CK) biosynthesis by activating the transcription of IPT7 gene, known for its crucial role in meristem maintenance. By applying exogenous CKs, the phenotype of *SMT* mutant can be reversed, and overexpression of SMT induces the biosynthesis of CKs and results in the formation of ectopic meristem [13–15]. Another gene of the KNOX gene family known as KNAT1/BREVIPEDICELLUS (BP) also regulates CK biosynthesis suggesting an active role for this gene in maintaining the meristem [14, 16]. Besides, SMT

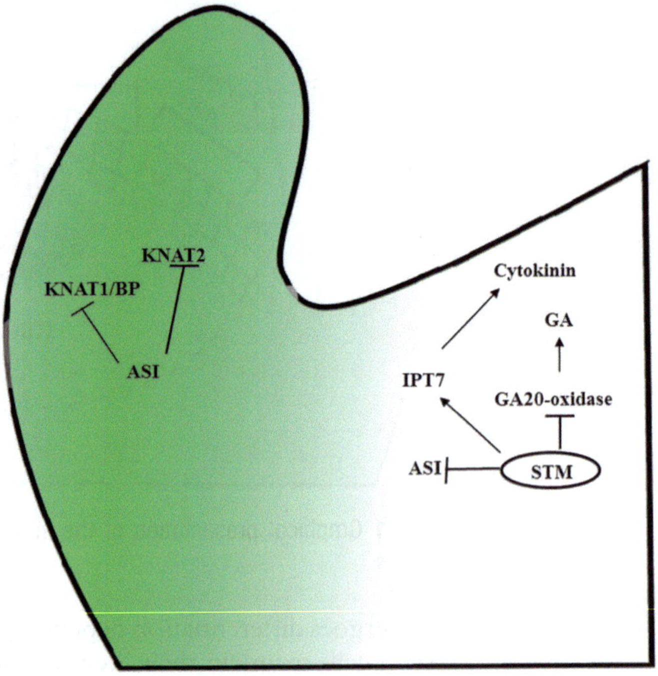

Fig. 2 Regulation of meristem boundaries by STM. STM promote CK biosynthesis and its activities by upregulating the IPT7 and downregulating AS1 which in association with auxins represses KNOX gene activities responsible for promoting the meristem. The STM is also known for its suppression of CKs and Gas in the region of leaf primordial

also represses the meristem-promoting activity of two factors including KNAT1/BP and KNAT2 in primordia by suppressing the expression of ASI gene in the shoot meristem [17]. Furthermore, STM downregulates the production of cell differentiation promoter gibberellic acid (GA) in the stem cell by suppressing GA 20-oxidase gene responsible for the biosynthesis of GA and upregulates GA 2-oxidase gene involved in degradation of active gibberellin (Fig. 2) [14, 18, 19].

1.3 WUS and ZLL Are Key Genes to Maintain the Stem Cell Meristem

An important gene known as *wuschel (WUS)* codes for a plant-specific homeodomain protein and is required for shoot meristem maintenance. WUS belongs to WUSCHEL-RELATED HOMEOBOX (WOX) gene family. Members of this family regulate diverse aspects of development [20]. Mutation in WUS gene leads to the absence of the meristem and partial differentiation of stem cells showing its importance in maintaining the stem cell undifferentiated [21, 22]. Overexpression of the WUS enlarges meristem, showing that this gene promotes stem cell identity [13, 23–25]. WUS gene is expressed in the organization center of the shoot meristem located underneath the three stem cell layers and maintains stem cells in two ways including its translocation

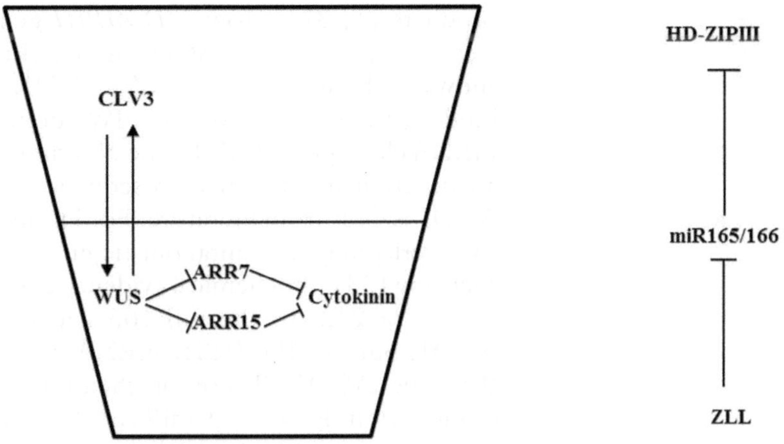

Fig. 3 Negative feedback loop of WUS/CLV3 maintains stem cell activity. WUS regulates ARR7/ARR15 expression to inhibit intracellular CK response and maintain stem cells. Expression of ZLL in the vasculature promotes the activities of HD-ZIPIII genes in the shoot primordium by sequestering miR165/miR166

(through plasmodesmata) from OC to the CZ where it binds to the promoter of CLV3 to initiate transcription. Decrease in the amount of the WUS leads to the loss of the shoot meristem suggesting that maintenance of stem cells is dependent on the translocation of WUS [9, 25, 26]. The WUS contribute to stem cell maintenance by regulating the expression of two type A *ARABIDOPSIS RESPONSE REGULATOR* genes ARR7 and ARR15 which encode the inhibitor of the intercellular response to CKs (Fig. 3). It suggests that the presence of WUS in both OC and stem cells is important to preserve the undifferentiated status of the stem cells. The results of chromatin immunoprecipitation and transcription profiling assay revealed that WUS act on vast majority of genes that regulate the meristem and control cell division and phytohormonal pathways [27, 28].

ZWILLE/PINHEAD/AGO10 (*ZLL*) encode a protein of 988 amino acids called ARGONAUTE (AGO) expressed in the vascular primordium playing a key role in the maintenance of meristem cells (Fig. 3). Mutation in this gene results in the differentiation of apical meristem stem cells [29, 30]. In *ZLL* mutant, *WUS* gene expressed normally in the OC, but the expression of CLV3 was not maintained. Overexpression of *WAS* gene in *ZLL* mutant also failed to accumulate stem cells in the meristem [31]. All these findings suggest that *ZLL* enhance *WUS*-dependent cell signaling. AGO proteins are the repressor of microRNA 165 and microRNA 166 which are the cleaver of the mRNA of *HOMEODOMAIN-LEUCINE ZIPPER III* (*HD-ZIPIII*) genes *ATHB-9/PHV*, *ATHB-14/PHB*, and *ATHB-15*. Mutation in the AGO-coding gene *ZLL* leads to the accumulation of microRNA 165 and microRNA 166 resulting in the loss of *HD-ZIPIII* gene products [1, 32]. How *HD-ZIPIII* gene products maintain the stem cell is not clear. AGO1 is the close homolog of ZLL and shows high similarity in PAZ and MID domains which help in binding to small RNAs and PIWI domain responsible for target mRNA cleavage in AGO1. The N terminal sequences of both genes are different and do not show sequence similarities. The dual role of AGO1 is clear from literature. For instance, this gene is responsible for overlapping and antagonistic effect on gene development and silencing [33]. Biochemical evidence demonstrates that the binding affinity of ZLL to the miR165 and miR166 is higher than the AGO1, but its HD-ZIPIII mRNA degradation efficiency is lower than the AGO1. These biochemical results suggest that ZLL sequesters miR165 and miR166 form the AGO1 to upregulate the expression of HD-ZIPIII [34].

1.4 Clavata and Its Receptors Contribute to OC Stability

Stem cell maintains their size, shapes, and internal organization, but how do they control the boundaries of meristem? Answer of this question became possible after observing extended stem pool and production of comparatively more organs in the *Clavata* (*clv*) mutant as compared to wild-type plants [35]. One of the family members of 32 small proteins (CLE family) called the CLV3 protein is important for intercellular talk [36]. The cause of the expanded meristem in the CLV3 mutant was the enlarged WAS domain [24]. Contrary, the overexpression of CLV3 represses the expression of WUS and shows phenocopy of *WUS* mutant [37]. The CLV3 expressed in the wedge-shaped domain coexist with stem cells. A WUS gene mutant *Arabidopsis* is characterized by the downregulation of CLV3, while its expression is upregulated in plants overexpressing WUS. This positive regulation of CLV3 by WUS constitutes a negative feedback loop where the CLV3 expression in the stem cells depends on the WAS expression in OC [24]. Thus, the stability and maintenance of stem cells in the SAM rest on this negative feedback loop (Fig. 3). For the homeostasis of stem cells in the SAM, perception of CLV3p by a receptor protein is a key step. Among these receptors, LRR receptor kinase encoded by CLV1 expresses in the central zone of SAM. Another gene, CLV2, encodes for a protein homolog of LRR receptor kinase that lakes intercellular kinase domain. This LRR receptor-like protein interacts with CORYNE (CRN)/SUPPRESSOR OF LLP1 2 (COL2) which has kinase domain but lakes the receptor domain. The third identified receptor in the CLV3p perception is RECEPTOR-LIKE PROTEIN KINASE 2 (RPK2)/TOAD-STOOL2 (TOAD2) [38, 39]. Triple mutant *clv1 clv2 rpk2 Arabidopsis* shows phenocopy of *clv3* mutant which suggests that these three receptors are involved in main pathways of CLV3p perception. CLV3 repress WUS at the distal and lateral boundaries of the organizing center (OC), and its overexpression in L1 layer leads to complete suppression of WUS [37]. The binding of the receptors to the CLV3p in overlying cells limits its spreading from the stem cell in lateral direction and maintains OC stability in the meristem [37]. Schoof et al. proposed that graded signal emanated from the stem cell promotes the expression of WUS keeping the stem cell niches at the tip of the plant [24]. Some observation suggests the involvement of CKs in this process [40].

2 Root Stem Cell Niches

The root tip has stem cells which are responsible for root growth toward the gravity. Central region of the root tip is called quiescent center (QC) which is made of relatively mitotically inactive cells. QC is surrounded by the stem cells and by a division which gives rise to different files of the root such as the epidermis, stele, endoderm, cortex, root cap, and columella (Fig. 4). An asymmetrical initial division of the root stem cell produces two cells, one of which remains in the QC and the other one undergoes several rounds of mitotic division and subsequent differentiation. Columella stem cells (CSCs) are located at the distal side of QC, and its daughter cells differentiate into starch-containing graviperceptive columella cells. Every stem cell of the root meristem has limited potential of differentiating into only one kind of tissue. Contrary to the lineage-based differentiation of the daughters of shoot stem cells, signals received from the already differentiated cells play a vital role in the differentiation of root stem cell daughters [1].

2.1 Organization of Stem Cell Niche in the Root

QC play a key role in controlling root stem cell function. The ablation of QC leads to the blockage of CSC proliferation and its differentiation to the starch-containing columella cells [41]. The QC signals are of short range as only the adjacent stem cells are maintained, and the nonadjacent stem cells differentiated. RETINOBLASTOMARELATED (RBR) also play a key role in maintaining stem cells as several layers of the cells remain undifferentiated around the QC in *Arabidopsis* where the expression of RBR has been downregulated by RNAi [42]. The undifferentiated state of these cell layers changes into differentiated cells when the QC was ablated which indicates that the stem cell-promoting signals were from QC and can work within some diameter, but normally they are counteracted by those cells which are not in direct contact with QC. The signals from the QC maintaining root stem cells are not yet discovered, but some important pathways are identified in recent works [1].

2.2 WOX5 Gene as Stem Cell Stabilizer

WOX5 is a homolog of WUS and expressed in the QC. The loss of the WOX5 function leads to the differentiation of CSCs same as in the result of ablation of QC. Overexpression of WOX5 represses the process of differentiation in the columella cells and makes stem

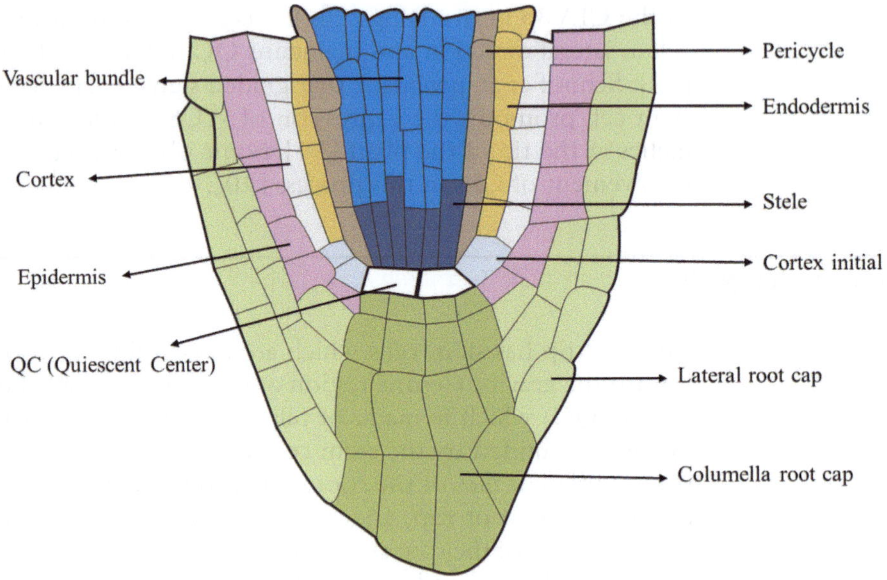

Fig. 4 Diagrammatic view of *Arabidopsis* root longitudinal section showing different regions of the root apical meristem in different colors

cell-like cells. The ablation of QC does not suppress the WOX5 overexpression effect on columella cells which indicates that there are no other signals except WOX5 from the QC which maintain the root stem cell in undifferentiated state [43].

Some CLE peptides play important role in differentiation and maintenance of shape and size of cells in the root meristem like CLE40 which is expressed in differentiated columella cells (Fig. 4) [36, 44]. CLE40 together with ACR4, which is expressed in the columella stem cells and first layer of differentiated columella cells, promote the cell differentiation of the columella cells by counteracting the stem cell-promoting signal from the QC cells [44].

2.3 SHR Signals From the Stele Are Important for QC Function

SHR is a GRAS (GAI, RGA, SCR) transcription factor and is expressed in the stele region. From the stele region, it moves to the neighbor cells including QC where they activate the expression of SCR gene which facilitates the nuclear localization of SHR (Fig. 5) [26]. Mutation in both genes, SHR and SCR, leads to the irregular morphology of root stem cell niches and downregulation of QC-specific markers and finally results in the collapse of the root meristem. In *shr* mutant, the QC-specific expression of SCR failed to rescue the QC defect of *shr* indicating that the presence of the products of these two genes is necessary in the QC for its stem cell maintenance activity [45–47].

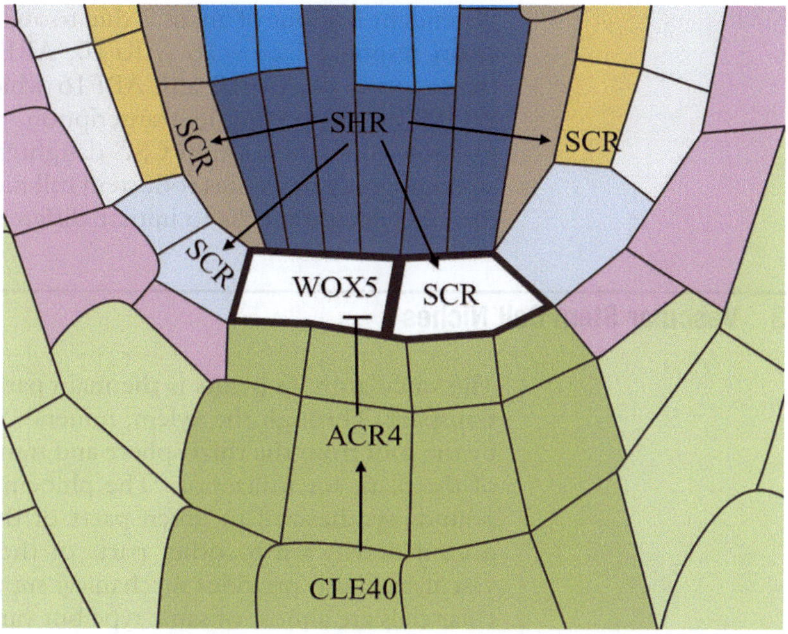

Fig. 5 The transcription factor SHR moves out of the stele region (expression spot) to the adjacent cells and QC where the expression of SCR is promoted resulting in nuclear localization of SHR. The stem cell-maintaining signals from QC are suppressed by CLE40 (expressed in differentiated columella cells) and ACR4 (expressed in the first layer of columella stem cells)

2.4 Stem Cell Maintenance of Auxin via PLT

Auxin accumulation in the QC is achieved by shootward auxin transport in the epidermis and lateral root cap and rootward auxin transport in the vascular tissues [48]. With the removal of the root tip, auxin starts to accumulate in epic cells and establish auxin maximum which leads to the formation of new root tip and new stem cell niches [49]. This indicates that auxin maximum and stem cell niches are functionally linked. Auxin maintains the stem cell niches via *PLT* transcription factors [50]. The activity of PLT in the cells depends on its expression level indicating its dose-dependent function. High level of PLT expression in the QC promotes stem cell niches, intermediate level of expression in the proximal meristem promotes mitotic cell division, and low level of expression initiates differentiation. Auxin is indirectly linked with PLT expression by tyrosylprotein sulfotransferase (TPST) and root growth factor (RGF). Auxin enhances the expression of TPST and RGF which results in TPST sulfate, the RGF proteins which upregulate the expression of PLT by an unknown mechanism. PLT proteins enhance the expression of PIN, creating a positive feedback loop for the stabilization of auxin maximum at root tips [51, 52]. The function of the auxin in the root meristem depends on cell texture, e.g., in QC, auxin helps in the maintenance of stem cells, while, in the columella, it promotes cell differentiation. This cell texture-dependent readout of auxin is due to auxin response factor 10 and auxin response factor 16 (ARF10, ARF16). Auxin activates the transcription of ARF10 and ARF16 which restrict the activity of WOX5 by suppressing its transcription, restrict it to the QC, and promote differentiation in CSC daughter cells [53, 54]. In short, auxin not only promotes root stem cell niches but also restricts it in the CSC daughter cells to initiate differentiation.

3 Vascular Stem Cell Niches

The vasculature in plants is the main path of mineral and nutrient transport. Through the xylem, minerals in the water are absorbed by the root from the rhizosphere and transported to the upper part of the plant for utilization. The phloem transports organic compounds synthesized in green parts of the plant which are transported downward to other parts of the plant. Besides that, the vasculature also provides mechanical support to the stem. All vascular cells are almost of same type but vary in architecture depending on its position. In *Arabidopsis* root, the centrally located metaxylem is sorrounded by protoxylem. On both sides of the xylem, the phloem is located at perpendicular axis to the xylem. The region between the xylem and phloem is occupied by the procambium which is consist of pluripotent stem cells. The xylem, procambium, and phloem are surrounded by pericycle and form the vasculature (Fig. 6a). In the stem, vascular bundles present in ring form where the procambium (fascicular cambium) presents at central position, the xylem on the inside, and the phloem on the outside [1]. During the secondary growth, the procambium is connected by the interfascicular cambium and forms a closed ring.

3.1 Stem Cell Maintenance in the Procambium

Similar to the shoot meristem, cambium stem cells are maintained by many key regulator genes, most of which are identified and show similarities to the apical meristem. Tracheary element differentiation inhibitory factor (*TDIF*) encodes CLE peptide protein, promotes cell proliferation, and represses differentiation in xylem cells (Fig. 7) [36]. *CLE41* and *CLE44* are homologs of *Arabidopsis TDIF* (tracheary element differentiation inhibitory factor) gene, are expressed in the phloem cells, and induce proliferation in the neighboring procambial cells of hypocotyl and shoot. PHLOEM INTERCALATED WITH XYLEM/TDIF RECEPTOR (PXY/TDR) receptor-like kinase are CLVI-like proteins, perceive CLE41 and CLE44 protein in the stem cells, and promote cell division. Lake of PXY/TDR results in reduced number of procambial calls, loss of procambium cell division orientation, and interspersed xylem and phloem [55, 56]. It indicates that differentiating phloem daughter cells maintain stem cell stability through stem cell-promoting signals and behave like niche cells similar to the OC and QC in the root and shoot meristem, respectively.

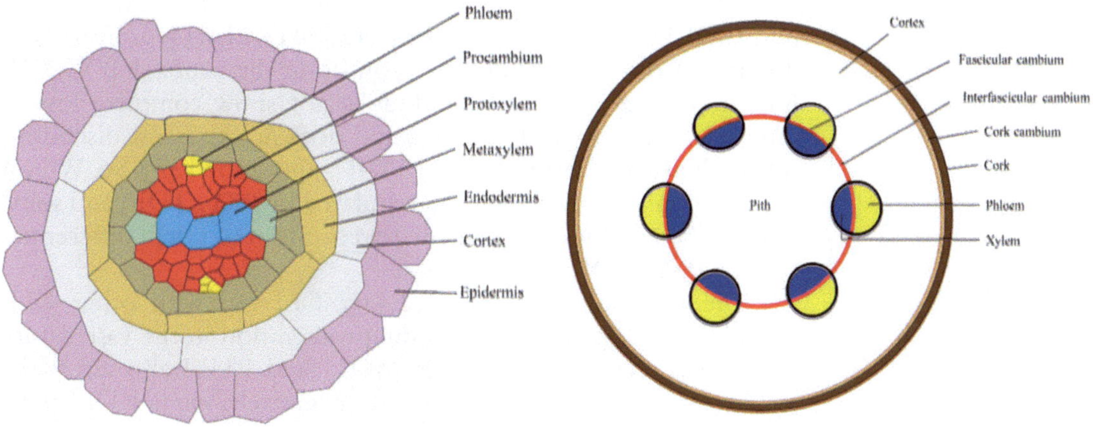

Fig. 6 Vascular patterning and tissue organization in *Arabidopsis* root (**a**) and stem (**b**)

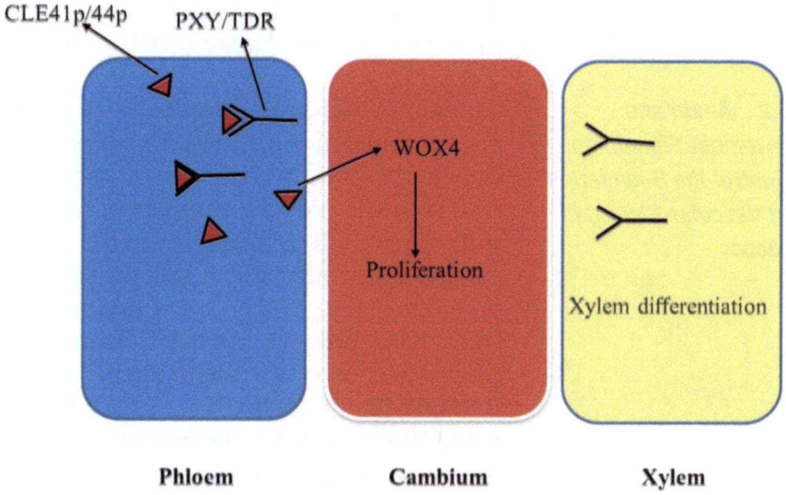

Fig. 7 The CLE–PXY/TDR–WOX4 pathway responsible for the proliferation of stem cells driving orientation of the cell division

Ubiquitous overexpression of CLE41/CLE44 represses xylem differentiation and accumulates calls in vascular bundles and interfascicular region, but the specific overexpression of CLE41 in phloem cells and ubiquitous overexpression of PXY/TDR cannot repress the differentiation of xylem cells. From this, it can be concluded that some other unknown factors limit the range of CLE41p by an unknown mechanism. Therefore, CLE41p-PXY/TDR module defines the boundaries between vascular cell types and also regulates the number of stem cells. CLE41 ubiquitous and xylem-specific expressions induce disorientated procambial cell division, but its expression in the phloem cells induces normal cell division. This indicates that the position of the stem cell's relative cells producing CLE peptide correlates with the orientation of stem cell division [55].

WOX4 is another gene in the procambium which expression is regulated by CLE41 and CLE44. CLE41/CLE44 promote cell division in the procambium by promoting the expression WOX4. Unlike pxy/*tdr*, loss of WOX4 does not show complete loss of intervening procambial cell layer and does not suppress discontinuous xylem stand formation and TDIF application [57]. So, WOX4 mediates only stem cell division regulated by PXY/TDR, and some unidentified pathway must mediate repression of xylem differentiation by PXY/TDR (Fig. 7).

The application of the CLE41p/CLE44p with other CLE peptides to the plant results in the proliferation in the vasculature [58]. Two receptor-like kinases MOLI and RULI also affect the activity of cambium. MOLI regulate negatively, while RULI positively regulate cambium activity [59]. Furthermore, the result of transcript profiling of *Arabidopsis* and *Populus* indicates that two shoot meristem regulators CLV1 and STM are also important for the maintenance of vascular stem cells, but the mechanism on how they regulate vascular stem cells is not clear [60, 61].

3.2 Auxin and Cytokinin Signalings Control the Boundaries of Vascular Stem Cell Niches

Cytokinins (CKs) translocate from the shoot to the root via symplastic connection of the phloem and maintain the status of stem cells in the procambium. Any reduction in the CKs or its signaling results in the reduced number of cells in the vasculature [1]. The site of active CK signaling lies in the procambial cells adjacent to the xylem axis which affect the localization of PIN3 (expressed in pericycle) and PIN7 (expressed in phloem and procambium) [62]. This bisymmetrical localization of PIN channels auxin to central axis of xylem where auxins induce the expression of ARABIDOPSIS HISTIDINE PHOSPHOTRANSFER PROTEIN 6 (AHP6) which blocks protoxylem formation by repressing CK signaling in the protoxylem position [62]. Thus, the mutational impeding interaction of CKs and auxins determines the boundaries between protoxylem and procambium stem cells (Fig. 8).

ZIPIII protein promotes xylem differentiation in procambial cells [63]. The transcription factors SCR and SHR activate the expression of miR156a/miR166b genes in the endoderm. From the endosperm, miR156a/miR166b move to the stele center where they suppress the expression of HD-ZIPIII genes *PHB*, *PHV*, *REV*, *CORONA (CNA)*, and *ARABIDOPSIS HOMEOBOX* GENE by degrading their transcripts. The concentration gradient of HD-ZIPIII protein determines development of xylem in a concentration-dependent way. High level of this protein produces xylem, while lower level produces protoxylem [64]. High level of this protein produce xylem while lower level produce protoxylem [64]. All these findings conclude that neighboring cells along the xylem axis are important because their positional signals decide the fate of the cells in their vicinity.

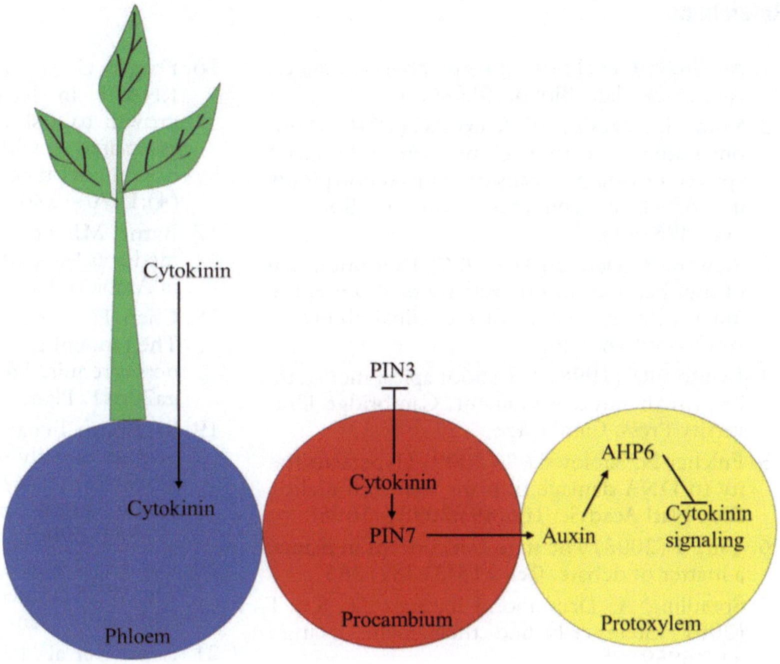

Fig. 8 Auxin and cytokinin signalings control the boundaries of vascular stem cell niches

3.3 JA Signaling Stimulates Interfascicular Cambium Initiation

During plant development, the plant gains weight specifically by producing green canopy. To bear increasing weight, the plant starts secondary growth in response to give strength to the stem. It has been proven from the experimental work that loading artificial weight on the immature plant tip induces IC formation, possibly through auxin signaling. But recent studies did not find any direct correlation between plant weight and IC initiation [65]. Recent studies demonstrate that JA signaling is involved in secondary growth. *JAZ10* is a touch-inducible JA signaling gene and is expressed in the xylem and IC of the basal stem. The intra-tissue tension which develops either as a result of cell division in the fascicular cambium or pushing of cambium outward due to the generation of xylem is thought to be important in the activation of JA signaling and subsequent initiation of IC formation. Hence, the intra-tissue tension and body weight are supposed to be involved in the stem cell niches of cambium [66].

References

1. Aichinger E et al (2012) Plant stem cell niches. Annu Rev Plant Biol 63:615–636
2. Satina S, Blakeslee AF, Avery AG (1940) Demonstration of the three germ layers in the shoot apex of Datura by means of induced polyploidy in periclinal chimeras. Am J Bot 27 (10):895–905
3. Stewart R, Dermen H (1970) Determination of number and mitotic activity of shoot apical initial cells by analysis of mericlinal chimeras. Am J Bot:816–826
4. Lyndon RF (1998) The shoot apical meristem: its growth and development. Cambridge University Press, Cambridge
5. Fulcher N, Sablowski R (2009) Hypersensitivity to DNA damage in plant stem cell niches. Proc Natl Acad Sci 106(49):20984–20988
6. Laux T (2003) The stem cell concept in plants: a matter of debate. Cell 113(3):281–283
7. Spradling A, Drummond-Barbosa D, Kai T (2001) Stem cells find their niche. Nature 414(6859):98
8. Hake S, Vollbrecht E, Freeling M (1989) Cloning Knotted, the dominant morphological mutant in maize using Ds2 as a transposon tag. EMBO J 8(1):15–22
9. Smith LG et al (1992) A dominant mutation in the maize homeobox gene, Knotted-1, causes its ectopic expression in leaf cells with altered fates. Development 116(1):21–30
10. Long JA et al (1996) A member of the KNOTTED class of homeodomain proteins encoded by the STM gene of Arabidopsis. Nature 379(6560):66
11. Endrizzi K et al (1996) The SHOOT MERISTEMLESS gene is required for maintenance of undifferentiated cells in Arabidopsis shoot and floral meristems and acts at a different regulatory level than the meristem genes WUSCHEL and ZWILLE. Plant J 10 (6):967–979
12. Long JA, Barton MK (1998) The development of apical embryonic pattern in Arabidopsis. Development 125(16):3027–3035
13. Brand U et al (2002) Regulation of CLV3 expression by two homeobox genes in Arabidopsis. Plant Physiol 129(2):565–575
14. Jasinski S et al (2005) KNOX action in Arabidopsis is mediated by coordinate regulation of cytokinin and gibberellin activities. Curr Biol 15(17):1560–1565
15. Yanai O et al (2005) Arabidopsis KNOXI proteins activate cytokinin biosynthesis. Curr Biol 15(17):1566–1571
16. Frugis G et al (2001) Overexpression of KNAT1 in lettuce shifts leaf determinate growth to a shoot-like indeterminate growth associated with an accumulation of isopentenyl-type cytokinins. Plant Physiol 126 (4):1370–1380
17. Byrne ME et al (2000) Asymmetric leaves1 mediates leaf patterning and stem cell function in Arabidopsis. Nature 408(6815):967
18. Chen H, Banerjee AK, Hannapel DJ (2004) The tandem complex of BEL and KNOX partners is required for transcriptional repression of ga20ox1. Plant J 38(2):276–284
19. Sakamoto T et al (2001) KNOX homeodomain protein directly suppresses the expression of a gibberellin biosynthetic gene in the tobacco shoot apical meristem. Genes Dev 15 (5):581–590
20. van der Graaff E, Laux T, Rensing SA (2009) The WUS homeobox-containing (WOX) protein family. Genome Biol 10(12):248
21. Laux T et al (1996) The WUSCHEL gene is required for shoot and floral meristem integrity in Arabidopsis. Development 122(1):87–96
22. Mayer KF et al (1998) Role of WUSCHEL in regulating stem cell fate in the Arabidopsis shoot meristem. Cell 95(6):805–815
23. Lenhard M, Jürgens G, Laux T (2002) The WUSCHEL and SHOOTMERISTEMLESS genes fulfil complementary roles in Arabidopsis shoot meristem regulation. Development 129 (13):3195–3206
24. Schoof H et al (2000) The stem cell population of Arabidopsis shoot meristems is maintained by a regulatory loop between the CLAVATA and WUSCHEL genes. Cell 100(6):635–644
25. Yadav RK, Tavakkoli M, Reddy GV (2010) WUSCHEL mediates stem cell homeostasis by regulating stem cell number and patterns of cell division and differentiation of stem cell progenitors. Development 137 (21):3581–3589
26. Nakajima K et al (2001) Intercellular movement of the putative transcription factor SHR in root patterning. Nature 413(6853):307
27. Leibfried A et al (2005) WUSCHEL controls meristem function by direct regulation of cytokinin-inducible response regulators. Nature 438(7071):1172
28. Busch W et al (2010) Transcriptional control of a plant stem cell niche. Dev Cell 18 (5):841–853
29. McConnell JR, Barton MK (1995) Effect of mutations in the PINHEAD gene of

Arabidopsis on the formation of shoot apical meristems. Genesis 16(4):358–366
30. Moussian B et al (1998) Role of the ZWILLE gene in the regulation of central shoot meristem cell fate during Arabidopsis embryogenesis. EMBO J 17(6):1799–1809
31. Tucker MR et al (2008) Vascular signalling mediated by ZWILLE potentiates WUSCHEL function during shoot meristem stem cell development in the Arabidopsis embryo. Development 135(17):2839–2843
32. Liu Q et al (2009) The ARGONAUTE10 gene modulates shoot apical meristem maintenance and establishment of leaf polarity by repressing miR165/166 in Arabidopsis. Plant J 58 (1):27–40
33. Mallory AC et al (2009) Redundant and specific roles of the ARGONAUTE proteins AGO1 and ZLL in development and small RNA-directed gene silencing. PLoS Genet 5 (9):e1000646
34. Zhu H et al (2011) Arabidopsis Argonaute10 specifically sequesters miR166/165 to regulate shoot apical meristem development. Cell 145 (2):242–256
35. Clark SE, Running MP, Meyerowitz EM (1995) CLAVATA3 is a specific regulator of shoot and floral meristem development affecting the same processes as CLAVATA1. Development 121(7):2057–2067
36. Ito Y et al (2006) Dodeca-CLE peptides as suppressors of plant stem cell differentiation. Science 313(5788):842–845
37. Lenhard M, Laux T (2003) Stem cell homeostasis in the Arabidopsis shoot meristem is regulated by intercellular movement of CLAVATA3 and its sequestration by CLAVATA1. Development 130(14):3163–3173
38. Katsir L et al (2011) Peptide signaling in plant development. Curr Biol 21(9):R356–R364
39. Kinoshita A et al (2010) RPK2 is an essential receptor-like kinase that transmits the CLV3 signal in Arabidopsis. Development 137 (22):3911–3920
40. Yoshida S, Mandel T, Kuhlemeier C (2011) Stem cell activation by light guides plant organogenesis. Genes Dev 25(13):1439–1450
41. van den Berg C et al (1997) Short-range control of cell differentiation in the Arabidopsis root meristem. Nature 390(6657):287
42. Wildwater M et al (2005) The RETINOBLASTOMA-RELATED gene regulates stem cell maintenance in Arabidopsis roots. Cell 123(7):1337–1349
43. Sarkar AK et al (2007) Conserved factors regulate signalling in Arabidopsis thaliana shoot and root stem cell organizers. Nature 446 (7137):811
44. Stahl Y et al (2009) A signaling module controlling the stem cell niche in Arabidopsis root meristems. Curr Biol 19(11):909–914
45. Sabatini S et al (2003) SCARECROW is involved in positioning the stem cell niche in the Arabidopsis root meristem. Genes Dev 17 (3):354–358
46. Levesque MP et al (2006) Whole-genome analysis of the SHORT-ROOT developmental pathway in Arabidopsis. PLoS Biol 4(5):e143
47. Cui H et al (2007) An evolutionarily conserved mechanism delimiting SHR movement defines a single layer of endodermis in plants. Science 316(5823):421–425
48. Barton M (2010) Twenty years on: the inner workings of the shoot apical meristem, a developmental dynamo. Dev Biol 341(1):95–113
49. Wiśniewska J et al (2006) Polar PIN localization directs auxin flow in plants. Science 312 (5775):883–883
50. Galinha C et al (2007) PLETHORA proteins as dose-dependent master regulators of Arabidopsis root development. Nature 449 (7165):1053
51. Matsuzaki Y et al (2010) Secreted peptide signals required for maintenance of root stem cell niche in Arabidopsis. Science 329 (5995):1065–1067
52. Blilou I et al (2005) The PIN auxin efflux facilitator network controls growth and patterning in Arabidopsis roots. Nature 433 (7021):39
53. Wang J-W et al (2005) Control of root cap formation by microRNA-targeted auxin response factors in Arabidopsis. Plant Cell 17 (8):2204–2216
54. Ding Z, Friml J (2010) Auxin regulates distal stem cell differentiation in Arabidopsis roots. Proc Natl Acad Sci 107(26):12046–12051
55. Etchells JP, Turner SR (2010) The PXY-CLE41 receptor ligand pair defines a multifunctional pathway that controls the rate and orientation of vascular cell division. Development 137(5):767–774
56. Hirakawa Y et al (2008) Non-cell-autonomous control of vascular stem cell fate by a CLE peptide/receptor system. Proc Natl Acad Sci 105(39):15208–15213
57. Hirakawa Y, Kondo Y, Fukuda H (2010) TDIF peptide signaling regulates vascular stem cell proliferation via the WOX4 homeobox gene in Arabidopsis. Plant Cell 22(8):2618–2629
58. Whitford R et al (2008) Plant CLE peptides from two distinct functional classes

synergistically induce division of vascular cells. Proc Natl Acad Sci 105(47):18625–18630

59. Agusti J et al (2011) Characterization of transcriptome remodeling during cambium formation identifies MOL1 and RUL1 as opposing regulators of secondary growth. PLoS Genet 7 (2):e1001312
60. Zhao C et al (2005) The xylem and phloem transcriptomes from secondary tissues of the Arabidopsis root-hypocotyl. Plant Physiol 138 (2):803–818
61. Schrader J et al (2004) Cambial meristem dormancy in trees involves extensive remodelling of the transcriptome. Plant J 40(2):173–187
62. Bishopp A et al (2011) A mutually inhibitory interaction between auxin and cytokinin specifies vascular pattern in roots. Curr Biol 21 (11):917–926
63. Ilegems M et al (2010) Interplay of auxin, KANADI and class III HD-ZIP transcription factors in vascular tissue formation. Development 137(6):975–984
64. Carlsbecker A et al (2010) Cell signalling by microRNA165/6 directs gene dose-dependent root cell fate. Nature 465(7296):316
65. Ko J-H et al (2004) Plant body weight-induced secondary growth in Arabidopsis and its transcription phenotype revealed by whole-transcriptome profiling. Plant Physiol 135 (2):1069–1083
66. Sehr EM et al (2010) Analysis of secondary growth in the Arabidopsis shoot reveals a positive role of jasmonate signalling in cambium formation. Plant J 63(5):811–822

Chapter 2

Bacterial Shoot Apical Meristem Inoculation Assay

Muhammad Naseem, Gökhan Gun, Ozge Osmanoglu, Fatima A. AlRemeithi, Jibran Iqbal, and Thomas Dandekar

Abstract

By virtue of their sessile nature, plants may not show the fight-and-flight response, but they are not devoid of protecting themselves from disease-causing agents, attack by herbivores, and damages that are caused by other environmental factors. Plants differentially protect their life-sustaining organs such as plant apexes from the attack by microbial pathogens. There are well-established methods to inoculate/infect various plant parts such as leaves, roots, and stems with various different pathogens. The plant shoot apical meristems (SAM) are a high-value plant target that provides niche to stem cell populations. These stem cells are instrumental in maintaining future plant progenies by giving birth to cells that culminate in flowers, leaves, and stems. There are hardly few protocols available that allow us to study immune dynamics of the plant stem cells as they are hindered by various layers of the SAM cell populations. Here, we describe a step-by-step method on how to inoculate the *Arabidopsis* SAM with model plant pathogen *Pseudomonas syringae* pv. tomato DC3000.

Key words Shoot apical meristem, Immunity, Stem cells, Infection, CLV3p

1 Introduction

In plants, the shoot apical meristem (SAM) is a collection of undifferentiated and rapidly diving cells for the future organogenesis of the leaf, stem, and flowers [1–3]. SAM is divided into three different zones, including central zone containing undifferentiated stem cells, peripheral zone containing rapidly dividing cells, and rib zone [1, 3, 4]. It has been proposed that high levels of cytokinin signaling promote cell expansion and preservation of meristem cells because the deficiency of cytokinins in the SAM culminates into low meristematic activities [1, 5]. The patchy distribution of cytokinins and auxin in SAM cell layers regulates signaling networks that are fine-tuned by the loops regulated by signaling proteins such as WUS and CLV3p as well as signaling components of hormonal pathways such as auxin and cytokinins in the SAM [1].

Muhammad Naseem and Thomas Dandekar (eds.), *Plant Stem Cells: Methods and Protocols*, Methods in Molecular Biology, vol. 2094, https://doi.org/10.1007/978-1-0716-0183-9_2, © Springer Science+Business Media, LLC, part of Springer Nature 2020

Pseudomonas syringae pv. tomato DC3000 (*Pst* DC3000) is a bacterial plant pathogen with a hemibiotrophic lifestyle-causing bacterial speck on tomato and *Arabidopsis* [6] and has become an important model in molecular studies of plant-pathogen interactions [7, 8]. Natural infection cycle of *Pst* DC3000 begins upon contact with susceptible host through surface wounds or natural plant openings, such as stomata, and can multiply in the apoplast [9, 10]. Studying the bacterial pathogenesis and plant defense responses has been of great interest in understanding immune networks in plants. In general mechanism of host-pathogen interactions, plants have surveillance machines that recognize bacterial elicitors as well as effectors to prevent invasion by switching growth mode into defense mode [7, 11, 12]. The plant shoot apexes are assumed to grow sterile [13]. A complex signaling network keeps the stem cell populations as a block of undifferentiated cells in the plant shoot apical meristem [1, 3]. Inside the SAM, there is a physical hindrance posed by various SAM cell layers that protect the stem cell populations from being attacked by disease-causing agents. If accessed by any pathogen, the stem cell signaling can be badly interfered by external cues that may alter the stem cell programming into aberrant plant growth. Studies showed that the stem cells in the SAM of *Arabidopsis* express and secrete the CLAVATA3 peptide (CLV3p) which activates immune signaling via the flagellin receptor kinase FLAGELLIN SENSING 2 (FLS2) [3, 13, 14]. Here, we describe a modified method of infecting the plant SAM with model pathogen *Pst* DC3000. This will allow the plant community to perform experiments for the assessment of immune networks mediated by stem cell populations in the shoot apical meristem.

2 Materials

1. Murashige and Skoog (MS) medium: Dissolve 4.4 g of powder in 1 L of distilled water, and autoclave at 120–80 °C for 15 min.
2. 6-Well plate (9.5 cm^2 surface area).
3. Incubator at constant temperature of 28 °C.
4. LEDs providing constant 50–65 μmol m^{-2} s^{-1} light (*see* Fig. 1).
5. Glycerol stocks of *Pst* DC3000 stored at −80 °C.
6. Luria Broth (LB) medium: Dissolve 20 g of powder in 1 L of distilled water, and autoclave at 120–80 °C for 15 min.
7. Antibiotic: rifampicin, prepare 50 mg/mL stock solution in DMSO, and store at −20 °C.
8. LB agar plates: LB medium plus 15.0 g agar per liter and 50 μg/mL rifampicin.

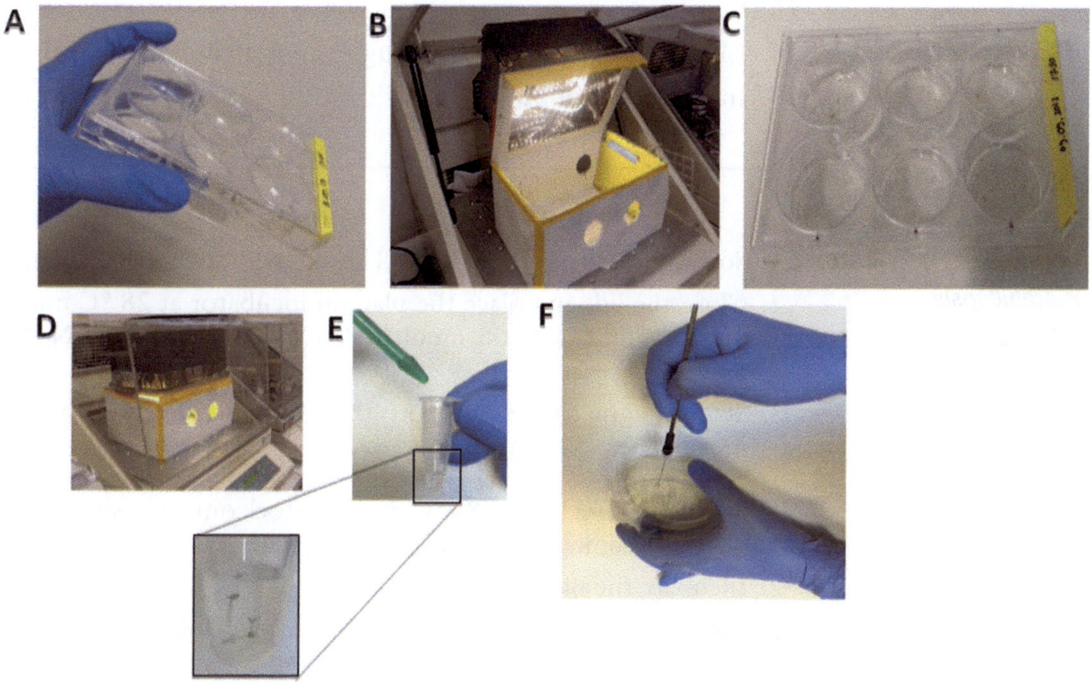

Fig. 1 Steps of SAM infection with *Pst* DC3000. *Arabidopsis thaliana* Col-0 seeds were cultured in 6-well plate (**a**) and inoculated into a growth chamber with 50 rpm shaking (**b**). Two-day-old seedlings and *Pst* DC3000 at OD600 0.02 were mixed (**c**), after 2, 3, and 4 days incubation (**d**), three seedlings were washed and grinded (**e**); number of colonies from infected seedlings were counted by plate-counting method (**f**)

9. Sterile 50 mL culture tubes for liquid bacterial growth.
10. Centrifuges.
11. Spectrophotometer to measure optical density of bacterial culture.
12. Sterile distilled water.
13. 2-day-old *Arabidopsis* seedlings (*see* Fig. 1).
14. 6-Well plate.
15. 0.5× Sterile MS medium.
16. Shaking incubator set at 28 °C.
17. Constant light source (50–65 μmol m^{-2} s^{-1}).
18. *Arabidopsis* seedlings inoculated with bacteria (*see* Fig. 1).
19. Sterile 1.5 mL microcentrifuge tubes.
20. Plastic micropestle proper for 1.5 mL microcentrifuge tubes.
21. Sterile distilled water.
22. 70% (v/v) Ethanol: H_2O.
23. Sterile toothpick or pipette tips.
24. Microcentrifuge.

25. Vortex.
26. LB agar plates containing 50 μg/mL of rifampicin.
27. Incubator set at 28 °C.

3 Methods

3.1 Growth of Arabidopsis Seedlings

1. Sow nine seeds in 1 mL 1× MS liquid medium in 6-well plates.
2. Close the lid, and place the plate in incubator at 28 °C under constant light (50–65 μmol m^{-2} s^{-1}) for 2 days (*see* **Notes 1** and **2**).
3. Set the light source far more than 20 cm from the plate to prevent evaporation.

3.2 Inoculation of Arabidopsis with Pst DC3000

1. Streak bacteria from −80 °C glycerol stock onto a fresh LB agar plate containing 50 μg/mL of rifampicin.
2. Incubate the plate at 28 °C for 24 h.
3. Inoculate 10 mL LB medium (with 50 μg/mL of rifampicin) with a single colony in a sterile tube.
4. Grow the bacterial culture at 28 °C at 200 rpm for 24 h.
5. Harvest bacteria by centrifugation at 3000 × *g* for 10 min.
6. Remove the supernatant, and resuspend the bacterial pellet with 10 mL sterile water, and centrifugate at 3000 × *g* for 10 min (repeat this step two times).
7. Dilute the bacteria to $OD_{600} = 0.02$ with distilled water.
8. Incubate the 50 μL of bacteria and 2-day-old seedlings into 1 mL fresh 0.5× MS medium in 6-well plate.
9. Grow the inoculated plate containing small seedlings at 28 °C with gentle shaking (50 rpm) for 2, 3, or 4 days under constant light.

3.3 Bacterial Quantification

1. Harvest the seedlings from infection plate 2, 3, and 4 days post inoculation with a sterile toothpick or pipette tips. Take three seedlings for each harvest (*see* **Note 1**).
2. Place the seedling into 1.5 mL tube, and wash three times followed by one wash with 1 mL 70% ethanol and two washes with distilled water. Harvest them by centrifugation for 60 s at 5000 × *g*. Grind the seedlings in the 1.5 mL tube containing 100 μL of distilled water using a hand drill and plastic micro-pestle until the seedlings are completely dissolved and no plant pieces are visible (*see* Fig. 1).
3. Add 900 μL of sterile distilled water to the sample and vortex vigorously.

4. Make a dilution by removing 100 μL of the main suspension, and add into a new tube containing 900 μL of distilled water.
5. Spread 50 μL of each sample (main solution and diluted suspension) onto LB agar plate with antibiotic (50 μg/mL rifampicin) in triplicate.
6. Incubate plates at 28 °C for approximately 2 days (*see* **Notes 3** and **4**) or until the colonies are visible (*see* Fig. 1).

4 Notes

1. To obtain three seedlings with nearly identical growth stage in single trial of inoculation, six to eight seeds should be added to each well in the beginning (*see* Fig. 1). Also, it is advisable that one or two extra well of the plate should be used in addition to the three well for triplicate to avoid any mishap specifically due to evaporation or poor handling of the culture.
2. Depending on the light source, keep the samples approximately 20–25 cm distance from the light source to avoid evaporation and warming of the solutions inside the plates. Plates have to be stable in the incubator at 50 rpm shaking by a basic sticker bottom of the plates.
3. Adjust the optical density of the bacteria using a spectrophotometer at 600 nm. To make sure the dilution set at $OD_{600} = 0.02$ (approximately 1×10^7 CFU/mL), each calculated serial dilution should be measured to be optimized in each laboratory setting. After setting the optical density at 0.02, 10 μL of bacteria has to be spread on agar plates (in triplicate) to verify suspension preparation (*see* Fig. 1).
4. Depending on the *Arabidopsis* lines, bacterial colony counts can be different significantly from each other due to differences in host defenses and disease progression. The amount of spreading of the harvested bacteria from seedlings can be optimized by serial dilutions or using more/less concentrated suspension.

Acknowledgment

We thank the German Research Foundation (DFG) for funding (TR124/B1) to TD and start-up grant (R18045) by Zayed University to MN.

References

1. Hwang I, Sheen J, Müller B (2012) Cytokinin signaling networks. Annu Rev Plant Biol 63:353–380. https://doi.org/10.1146/annurev-arplant-042811-105503
2. Aichinger E, Kornet N, Friedrich T, Laux T (2012) Plant stem cell niches. Annu Rev Plant Biol 63:615–636. https://doi.org/10.1146/annurev-arplant-042811-105555
3. Naseem M, Srivastava M, Dandekar T (2014) Stem-cell-triggered immunity safeguards cytokinin enriched plant shoot apexes from pathogen infection. Front Plant Sci. https://doi.org/10.3389/fpls.2014.00588
4. Miwa H, Betsuyaku S, Iwamoto K et al (2008) The receptor-like kinase SOL2 mediates CLE signaling in arabidopsis. Plant Cell Physiol 49:1752–1757. https://doi.org/10.1093/pcp/pcn148
5. Sablowski R (2011) Plant stem cell niches: from signalling to execution. Curr Opin Plant Biol 14:4–9. https://doi.org/10.1016/j.pbi.2010.08.001
6. Cuppels DA (1986) Generation and characterization of Tn5 insertion mutations in pseudomonas syringae pv. Tomato. Appl Environ Microbiol 51:323–327
7. Jones JDG, Dangl JL (2006) The plant immune system. Nature 444:323–329. https://doi.org/10.1038/nature05286
8. Rico A, Preston GM (2008) Pseudomonas syringae pv. Tomato DC3000 uses constitutive and apoplast-induced nutrient assimilation pathways to catabolize nutrients that are abundant in the tomato apoplast. Mol Plant-Microbe Interact 21:269–282. https://doi.org/10.1094/MPMI-21-2-0269
9. Ritchie (2000) Bacterial spot of pepper and tomato. The Plant Health Instructor. https://doi.org/10.1094/PHI-I-2000-1027-01
10. Naseem M, Shams S, Roitsch T (2017) Modulating the levels of plant hormone cytokinins at the host-pathogen interface. Methods Mol Biol 1569:141–150
11. Naseem M, Kaltdorf M, Dandekar T (2015) The nexus between growth and defence signalling: auxin and cytokinin modulate plant immune response pathways. J Exp Bot 16:4885–4896
12. Newman M-A, Sundelin T, Nielsen JT, Erbs G (2013) MAMP (microbe-associated molecular pattern) triggered immunity in plants. Front Plant Sci 4:139. https://doi.org/10.3389/fpls.2013.00139
13. Lee H, Chah OK, Sheen J (2011) Stem-cell-triggered immunity through CLV3p-FLS2 signalling. Nature 473:376–379
14. Boller T, Felix G (2009) A renaissance of elicitors: perception of microbe-associated molecular patterns and danger signals by pattern-recognition receptors. Annu Rev Plant Biol 60:379–406. https://doi.org/10.1146/annurev.arplant.57.032905.105346

Chapter 3

Germline-Transmitted Genome Editing Methodology in *Arabidopsis thaliana* Using TAL Effector Nucleases

Joachim Forner

Abstract

TAL effector nucleases (TALENs) are powerful tools to create specific knockout mutants in plants. The use of an optimized TALEN backbone and the choice of promoters that are strongly active in the stem cells of the shoot apical meristem are key to a high rate of heritable targeted mutations. Recommendations for construct design and screening for mutants are given in this chapter.

Key words Genome editing, TALEN, Germline, *Arabidopsis thaliana*, Heritable, Knockout, NHEJ, Stem cell-specific promoter

1 Introduction

Transcription activator-like effector nucleases (TALENs) are engineered, site-specific DNA endonucleases [1]. They are based on TAL effectors (TALEs) which certain plant pathogens, such as *Xanthomonas*, excrete to activate transcription of specific genes in the nuclei of the host cells [2]. The DNA-binding domain of these proteins consists of repeats of 34 amino acids which vary only at positions 12 and 13, known as the repeat-variable di-residues (RVDs). One repeat binds one nucleotide on the DNA, and the amino acid combination at the RVDs specifies to which base. Repeats specific for adenine (A), cytosine (C), guanine (G), and thymine (T) can be freely combined to target any DNA sequence of choice. In TALENs, the transcription activation domain of the TALE proteins is replaced by a C-terminal FokI endonuclease domain [3]. Since this endonuclease domain can only function as a dimer, TALENs always need to work in pairs, binding DNA in a tail-to-tail orientation. DNA cleavage by TALENs results in double-strand breaks with staggered single-strand overhangs. In plant cell nuclei, these cuts are usually repaired via nonhomologous end joining (NHEJ) which creates small deletions or insertions

Muhammad Naseem and Thomas Dandekar (eds.), *Plant Stem Cells: Methods and Protocols*, Methods in Molecular Biology, vol. 2094, https://doi.org/10.1007/978-1-0716-0183-9_3, © Springer Science+Business Media, LLC, part of Springer Nature 2020

[4]. Thus, TALEN expression is a strong, site-specific mutagen. To make these mutations heritable, they need to take place in the germline, which can be achieved by driving TALEN expression by strong, stem cell-specific promoters.

The following chapter will provide some guidelines on how to use TALENs to create heritable loss-of-function mutations in arbitrary genes in *Arabidopsis thaliana*.

2 Materials

1. Molecular biology lab space.
2. Molecular biology typical equipment and consumables.
3. Plant cultivation and tissue culture facility.
4. Bacterial growth chambers and the basal techniques of microbial nature.
5. DNA and RNA extraction facility.
6. PCR and cloning facility.

3 Methods

3.1 General Considerations on Target Site Selection

1. When working with a single pair of TALENs, the resulting mutations will, most likely, be small deletions. Unless the cutting site encompasses a codon for an amino acid essential for the function of the targeted protein, the desired knockout effect will most likely stem from the frameshift occurring in two out of three cases of error-prone repair.
2. The more of the coding sequence (CDS) that lies downstream of the cutting site, the more likely the truncated protein will be fully dysfunctional. So, unless enough is known about the residues essential for the function of the protein encoded by the target gene to select essential codons for cutting, the binding sites for the TALENs should be chosen in the beginning of the coding sequence. Due to the effective length of the TALEN binding sites (usually 2×17 bases $= 34$ bases), off-target sites are not an issue unless working with members of a gene family or otherwise duplicated sequence.
3. When two TALEN pairs are available for the same gene, a likely outcome of co-expression is the deletion of the complete sequence between the two cutting sites. In this case, it is advisable to place the two cutting sites in such a way as to remove the entire gene (at least the complete protein-coding part), from the genome, to ensure a complete knockout of gene function (*see also* **Note 1**).

4. Similar considerations hold true when not targeting protein-coding sequences but promoters or miRNA genes.

3.2 Assembly of the TALEN Coding Sequences

1. Since TALEN encoding sequences are necessarily highly repetitive and therefore very challenging for cloning (*see also* **Note 2**), assembling them oneself is not advisable unless working in a lab with a lot of expertise in this field. Easier options are to purchase them from a commercial supplier or to collaborate with a lab specializing in TALEN construction.
2. The expertise provided by such partners will also help with selecting the ideal precise target sites and prescreening for TALEN protein activity in, e.g., yeast. For those willing to create the TALEN coding sequences themselves, there are several kits available (e.g., [5, 6] or see https://www.addgene.org/talen/).

3.3 Choice of Promoters and TALEN Backbone

1. To ensure the frequent and reliable creation of knockout alleles that are passed on to the next generation, the TALENs must be strongly expressed in the L2 stem cells of the shoot apical meristem or in the gametes or their immediate precursors. In our experience, the ribosomal protein P16 (At3g60245) [7] (GenBank: KP293940.1) and the HMG box protein (At1g76110) [8] (GenBank: KP293940.1) promoters are ideally suited for this purpose. A TALEN pair driven by these two promoters yielded, based on phenotype, more than 50% knockout plants in the T1 generation with a heritability of up to 100% [9].
2. As for the TALEN backbone, we obtained our best results working with the setup depicted in Fig. 1 [9]. In short, the 15 newly assembled DNA-binding repeats replace the original TALE DNA-binding domain, except for the very first and the very last repeats. Furthermore, the TALENs differ from the original TALEs by a truncation of 153 and 238 amino acids at the N- and the C-terminus, respectively. Finally, a nuclear localization signal (NLS) and a FokI endonuclease domain are added to the N- and C-terminus, respectively.
3. For convenience, to reduce the number of constructs or the length of the T-DNA, the CDSs for both TALENs of a pair can be joined by the sequence coding for a 2A peptide, e.g., T2A [10]. Thus, a single open reading frame on a single mRNA from a single expression cassette will produce two separate polypeptides.

3.4 Cloning the Expression Constructs

1. To be functional in plants, the TALEN CDSs need to be combined with plant promoter and terminator sequences as well as a selection marker cassette for subsequent plant transformation. For floral dip transformation, the vector backbone

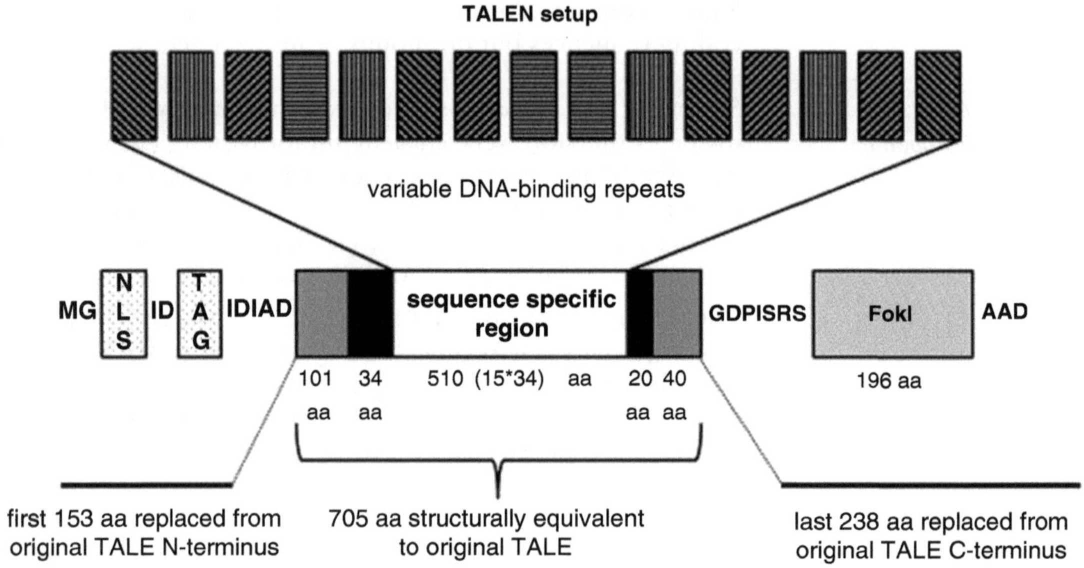

Fig. 1 Structure of a TALEN monomer. The variable DNA-binding repeats are depicted as hatched boxes, the FokI endonuclease domain as light gray box (FokI). Dark gray boxes represent unaltered regions also present in the original TAL effector, i.e., the truncated N- and C-terminus. The first, invariable thymine-binding repeat is shown as a black box, as is the last half-repeat. The thick black lines mark deleted parts of the original TAL effector. aa: amino acids. NLS: nuclear localization signal (DPKKKRKV). TAG: either HA-tag (YPYDVPDYA) or S-tag (KETAAAKFERQHMDS). For exact amino acid sequences, see, e.g., GenBank entry KP293938.1. Sequence of variable DNA-binding repeats: LTP Q/E QVVAIASxxGGKQALETVQRLLPVLCQAHG; xx: HD (for cytosine), NG (for thymine), NI (for adenine), NN (for guanine). Total length of TALEN: 936 (with HA-tag) or 942 (with S-tag) amino acids. Not drawn to scale

must be able to also replicate in *Agrobacterium tumefaciens* and provide left and right border sequences delimiting the transferred DNA region. In principal, any standard cloning method for assembling the expression constructs is applicable.

2. A convenient cloning system is GreenGate [11] which we have successfully employed for work with TALENs [9]. If the user intends to express each single TALEN individually, the TALEN CDSs should be converted into GreenGate S-modules, i.e., C-modules with a stop codon. If the user intends to join the TALEN CDSs by a 2A peptide provided by a C-module, one of the TALENs of each pair should be converted into a GreenGate B-module and the other into a D-module. GreenGate cloning enables the combination of up to two TALEN pairs on a single construct when using supermodules with two 2A-fused TALEN CDSs each.

3.5 Plant Transformation

1. Any transformation procedure established for *Arabidopsis thaliana* can be employed, the easiest being the floral dip transformation protocol [12]. If more than one construct needs to be transferred into a single plant, an alternative to

crossing of independently transformed plants might be co-transformation which has an efficiency of up to 30% when compared to single transformation [13]. In this case, two different selection markers have to be used for the individual constructs. *See* also Subheading 3.8 for remarks on the choice of resistance genes.

3.6 Screening for Mutants

1. After the T1 seeds have been harvested and sown on selection media to select for the presence of the TALEN transgene(s), the resulting resistant seedlings can be screened for TALEN activity, i.e., the generation of knockout alleles. If the mutation creates an easily visible phenotype, it suffices to check the T1 plants by eye and harvest seeds from plants that are either fully or partially affected. In the latter case, seeds should be collected only from those parts of the plants that display the mutant phenotype.
2. If no obvious macroscopic effect of the mutation is to be expected, the individual plants need to be genotyped by PCR (*see also* **Note 3**). DNA should be isolated from reproductive tissue, preferably pooled from several parts of an individual plant. When two TALEN pairs are used in parallel, the resulting large deletion is reflected in a size difference between wild-type and deletion allele PCR products that is easily detectable. The relatively small deletions and insertions resulting from a single cut, however, are usually not sufficient to result in PCR product sizes clearly different from the wild-type allele. In this case, the PCR products should be either sequenced directly or subcloned and several individual clones analyzed.
3. Direct sequencing will show whether or not non-wild-type alleles are present at all, while subcloning will provide additional information about the nature of the alternative alleles.

3.7 Confirmation of the Mutant Genotype/Phenotype in the Next Generation

1. Having confirmed that a certain plant contains no, or very few, wild-type alleles of the target locus, it is still necessary to check the genotype and/or phenotype of its offspring in the next generation (*see also* **Note 2**). The parental plant could have been chimeric, meaning that small sectors of non-mutated cells might have given rise to wild-type offspring. Also, only a minority of the alleles in somatic tissue are derived from the L2 stem cells, but they are the exclusive source of the gametes, so there might be a discrepancy between the genotyping results in the parent and in the offspring.
2. Furthermore, even a plant with an unambiguous knockout phenotype might possess two different null alleles of the target gene (i.e., being biallelic), so that it might be advisable to harvest seeds from individual plants of the next generation to have genetically uniform material.

3.8 Eliminating the TALEN Constructs from the Genome

1. To avoid any possible future off-target effects and to ensure a true Mendelian segregation of the knockout phenotype in crossings, it is preferable to work with descendant plants that have inherited two copies of a null allele of the target gene, but no copy of the TALEN transgene. Such plants can be very easily screened for if the chosen marker gene allows not only positive selection to identify transgenic T1 plants but also negative selection for its absence in later generations. Such selection markers include, for example, the *codA:nptII* fusion [14] or the *dao1* gene [15]. This latter encodes a D-amino acid oxidase which detoxifies D-alanine so that only transgenic plants can survive on medium containing D-alanine (at >3 mM), which allows positive selection for the transgene.
2. But it also converts the harmless D-valine into a toxic keto acid (3-methyl-2-oxobutanoate), which enables counter-selection against the presence of the transgene on media with D-valine (>15 mM). For this purpose, seedlings from either selfing of an engineered knockout plant or the F2 from outcrossing with wild type are grown on D-valine and the survivors then genotyped for mutations in the target gene.

4 Notes

1. Target site selection

   The first base recognized by any TALEN is a T. The following 16 repeat units can be chosen freely. The length of the spacer between the TALEN binding sites depends on the selected backbone; for the one recommended here, it is 14 base pairs. Thus, TALENs can in principle be engineered for any T-46xN-A or A-46xN-T motif (i.e., at roughly every eighth position).

   The most N-terminal 34 amino acid repeat of the TALEN binds the most 5′ nucleotide of the binding site and the most C-terminal repeat the most 3′ nucleotide. The cut occurs approximately at the midpoint between the two TALEN binding sites (*see* Fig. 2).

   When more than two pairs of TALENs are used simultaneously against the same gene, their binding sites should be chosen in a way so as to ensure that any combination of two cuts will both very likely create a knockout allele and a deletion large enough to easily detect via PCR. For example, if the first two pairs frame the entire CDS, the third one should be placed around the middle of the reading frame or slightly in front of the middle.

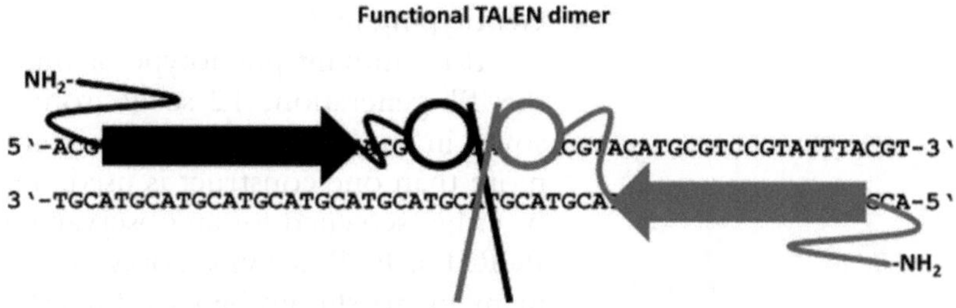

Fig. 2 Principle of TALEN function. Two different TALEN monomers (marked in black and dark gray, respectively) bind adjacent DNA sequences in tail-to-tail orientation. The FokI endonuclease domains form a functional dimer at their C-termini. Only then can a DNA double-strand break be created. Arrows: DNA-binding parts of the TALENs, composed of the individual 34-mer repeats. Scissor halves: FokI endonuclease domain. NH_2: N-terminus

2. Cloning

   Due to the highly repetitive nature of the TALEN DNA-binding domain, it is extremely challenging to amplify the respective part of the CDS via PCR. Thus, cloning strategies should not include PCR steps. This also means that the TALEN CDS should be assembled in a vector that provides unique restriction enzyme cutting sites in close vicinity to the TALEN start and stop codon, regardless of the downstream cloning method (ligation independent or classical).

3. Genotyping

   Genotyping by PCR can be simplified if the TALEN cutting site coincides with a restriction enzyme cutting site. The latter is usually destroyed upon NHEJ repair of the TALEN cut and its presence thus indicative of the wild-type allele. When PCR products encompassing the putative TALEN cut site are digested with the respective restriction enzyme, only PCR products derived from the wild-type allele will be cut. Thus, the degree to which the PCR product can be cleaved is a direct readout for the percentage of wild-type alleles in the sample. Furthermore, should the TALEN activity be very low, PCR products from mutant alleles can be enriched by digesting the pool of products from the first PCR and using the cleavage-resistant part as template for a second PCR.

   When two TALEN pairs for the same locus are used simultaneously, it is possible, in rare instances, that the excised sequence can be reintegrated in the inverse orientation instead of generating a deletion. Genotyping for large size differences in these cases would incorrectly indicate the presence of a wild-type allele. Similarly, two subsequent cuts can occur, each sealed by NHEJ, thus possibly resulting in a mutant allele.

   As a rule of thumb, stop analyzing the T1 generation after 20–50 independent lines if no mutation is found, and continue with T2 (*see* **Note 4**).

4. Genotyping in T2

If no mutant phenotype or mutant alleles are detected in the T1 generation, T2 seeds from every T1 plant should be sown in bulk, preferably on selection medium, especially if more than one construct is used. These plants can then again be either screened for an observable phenotype or analyzed by PCR. For PCR analysis, pools from the reproductive tissues of many plants should be used for DNA isolation. For strategies to improve the ease of detection, *see* **Note 3**. In case no mutant phenotype/allele is detected in T2, the project should be stopped at that point. Although it cannot be ruled out that respective alleles might be detected in T3 or further generations, this is very unlikely, and thus no further effort is justified.

References

1. Christian M, Cermak T, Doyle EL, Schmidt C, Zhang F, Hummel A, Bogdanove AJ, Voytas DF (2010) Targeting DNA double-strand breaks with TAL effector nucleases. Genetics 186:757–761
2. Boch J, Scholze H, Schornack S, Landgraf A, Hahn S, Kay S, Lahaye T, Nickstadt A, Bonas A (2009) Breaking the code of DNA binding specificity of TAL-type III effectors. Science 5959:1509–1512
3. Pernstich C, Halford SE (2012) Illuminating the reaction pathway of the FokI restriction endonuclease by fluorescence resonance energy transfer. Nucleic Acids Res 3:1203–1213
4. Puchta H (2005) The repair of double-strand breaks in plants: mechanisms and consequences for genome evolution. J Exp Bot 409:1–14
5. Morbitzer R, Elsaesser J, Hausner J, Lahaye T (2011) Assembly of custom TALE-type DNA binding domains by modular cloning. Nucleic Acids Res 39:5790–5799
6. Cermak T, Doyle EL, Christian M, Wang L, Zhang Y, Schmidt C, Baller JA, Somia NV, Bogdanove AJ, Voytas DF (2011) Efficient design and assembly of custom TALEN and other TAL effector-based constructs for DNA targeting. Nucleic Acids Res 12:e82
7. Schuster C, Gaillochet C, Medzihradszky A, Busch AW, Daum G, Krebs M, Kehle A, Lohmann JU (2014) A regulatory framework for shoot stem cell control integrating metabolic, transcriptional, and phytohormone signals. Dev Cell 28:438–449
8. Yadav RK, Girke T, Pasala S, Xie M, Reddy GV (2009) Gene expression map of the Arabidopsis shoot apical meristem stem cell niche. Proc Natl Acad Sci U S A 106:4941–4946
9. Forner J, Pfeiffer A, Langenecker T, Manavella PA, Lohmann JU (2015) Germline-transmitted genome editing in Arabidopsis thaliana using TAL-effector-nucleases. PLoS One 10:e0121056
10. Daniels RW, Rossano AJ, Macleod GT, Ganetzky (2014) Expression of multiple transgenes from a single construct using viral 2A peptides in Drosophila. PLoS One 9:e100637
11. Lampropoulos A, Sutikovic Z, Wenzl C, Maegele I, Lohmann JU, Forner J (2013) GreenGate---a novel, versatile, and efficient cloning system for plant transgenesis. PLoS One 8:e83043
12. Clough SJ, Bent AF (1988) Floral dip: a simplified method for agrobacterium-mediated transformation of Arabidopsis thaliana. Plant J 16:735–743
13. De Buck S, Podevin N, Nolf J, Jacobs A, Depicker A (2009) The T-DNA integration pattern in Arabidopsis transformants is highly determined by the transformed target cell. Plant J 60:134–145
14. Schaart JG, Krens FA, Pelgrom KT, Mendes O, GJl R (2004) Effective production of marker-free transgenic strawberry plants using inducible site-specific recombination and a bifunctional selectable marker gene. Plant Biotechnol J 2:233–240
15. Erikson O, Hertzberg M, Nasholm T (2004) A conditional marker gene allowing both positive and negative selection in plants. Nat Biotechnol 22:455–458

Chapter 4

Method to Study Gene Expression Patterns During De Novo Root Regeneration from *Arabidopsis* Leaf Explants

Jie Yu, Ning Zhai, Lin Xu, and Wu Liu

Abstract

De novo root regeneration (DNRR) is the process in which adventitious roots are regenerated from damaged plant tissues or organs. We have developed a simple DNRR system in which adventitious roots are formed from detached leaf explants of *Arabidopsis* (*Arabidopsis thaliana*) on B5 medium without external hormones. In this chapter, we introduce the methods used to observe gene expression patterns during rooting from leaf explants. Usually, β-glucuronidase (GUS) staining is used to visualize gene expression patterns, since fluorescent proteins are difficult to observe because of the high autofluorescence in leaf explants. Here, we describe the use of the ClearSee technique with Congo red staining for deep imaging to observe fluorescent proteins. This method diminishes autofluorescence in leaf explants and preserves the stability of fluorescent proteins, thus allowing us to investigate the endogenous molecular actions guiding DNRR.

Key words De novo root regeneration, Gene expression pattern, GUS staining, ClearSee, Congo red staining, Plant regeneration, Adventitious root

1 Introduction

Plants have evolved various adaptive abilities to survive under severe environmental conditions. De novo root regeneration (DNRR) is a type of regeneration in which adventitious roots form from damaged plant bodies or detached explants [1–9]. Using leaf explants of *Arabidopsis* (*Arabidopsis thaliana*), we have developed a simple experimental system to study DNRR [1, 7]. In this system, the first pair of rosette leaves from 12-day-old seedlings is cut and then cultured on B5 medium without external hormones, and adventitious roots can be observed at 6–8 days after culture (DAC) (Fig. 1).

To understand the mechanisms that underlie DNRR, it is essential to visualize gene expression patterns at the cellular level.

Authors "Jie Yu" and "Ning Zhai" contributed equally to this work.

Muhammad Naseem and Thomas Dandekar (eds.), *Plant Stem Cells: Methods and Protocols*, Methods in Molecular Biology, vol. 2094, https://doi.org/10.1007/978-1-0716-0183-9_4, © Springer Science+Business Media, LLC, part of Springer Nature 2020

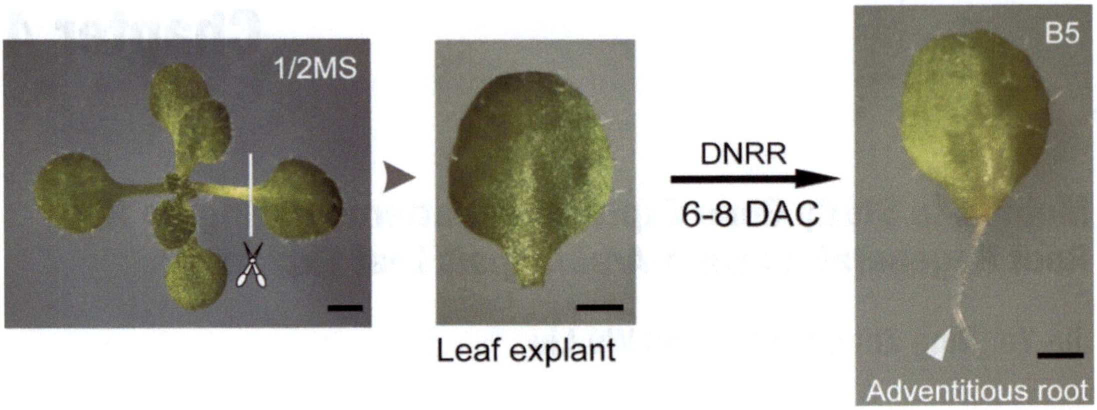

Fig. 1 De novo root regeneration from *Arabidopsis* leaf explant. Scale bars, 1 mm

Reporter genes such as those encoding green fluorescent protein (GFP) or β-glucuronidase (GUS) have been used extensively to monitor in situ gene expression in plant tissues [10, 11]. We can easily visualize gene expression during the regeneration process using GUS staining combined with differential interference contrast (DIC) microscopy at the tissue level. Alternatively, live imaging of proteins labeled with fluorescent tags can be observed by confocal laser scanning microscopy (CLSM). Live imaging of fluorescence-tagged proteins is more useful for the precise visualization and analysis of gene expression, since GUS staining by DIC observation without sectioning cannot detect gene expression in a single cell layer nor can multiple gene expression patterns be observed by multicolor imaging [12]. However, it is difficult to observe fluorescent proteins in leaf explants because they contain a variety of autofluorescent compounds, which result in nonspecific background fluorescence [12–14]. The transmission of fluorescence signals from deep leaf cells is also impeded by the heterogeneous refractive indices of the cell wall and cytoplasm [10, 12, 14–16]. Chloral hydrate-based reagents can be used to clear the tissues in leaf explants before DIC observations, but chloral hydrate also quenches the target fluorescent proteins. Recently, Kurihara et al. developed an aqueous chemical reagent, ClearSee, for deep imaging of fluorescent proteins in plant tissues. This reagent can diminish chlorophyll autofluorescence while maintaining the stability of fluorescent proteins [12]. We introduced this technique to our DNRR system and improved the method by adding a Congo red staining step to specifically mark the cell wall. This modified ClearSee method has allowed us to better observe gene expression patterns during the DNRR process by live imaging of proteins labeled with fluorescent tags.

2 Materials

2.1 Plant Materials

Arabidopsis thaliana Col-0 was used as the wild type. The *WOX5*$_{pro}$*:GUS* [17] and *WOX5*$_{pro}$*:eGFP* [18] marker lines were described previously.

2.2 In Vitro Culture Media

1. Half-strength Murashige and Skoog basal medium (½ MS medium): 2.2 g/L Murashige and Skoog basal medium powder, 1% (w/v) sucrose, 1% (w/v) agar, and 0.5 g/L 2-(N-morpholino)ethanesulfonic (MES) acid, pH 5.7 [19].
2. B5 medium: 3.21 g/L Gamborg's B5 basal medium, 0.5 g/L MES, 3% (w/v) sucrose, and 0.8% (w/v) agar, pH 5.7 [20].

2.3 Solutions for GUS Staining

GUS staining solutions are modified from the previous protocol [21].

1. 100 mM $K_4Fe(CN)_6$: dissolve 4.22 g $K_4Fe(CN)_6 \cdot 3H_2O$ in 100 mL H_2O.
2. 100 mM $K_3Fe(CN)_6$: dissolve 3.29 g $K_3Fe(CN)_6$ in 100 mL H_2O.
3. 1 M Na_2HPO_4: dissolve 179.11 g $Na_2HPO_4 \cdot 7H_2O$ in 500 mL H_2O.
4. 1 M NaH_2PO_4: dissolve 78.01 g $NaH_2PO_4 \cdot 2H_2O$ in 500 mL H_2O.
5. 0.5 M Na_2EDTA: dissolve 18.61 g Na_2EDTA in 100 mL H_2O; adjust pH to 8.0.
6. GUS staining solution (50 mM sodium phosphate buffer pH 7.0, 5 mM Na_2EDTA, 2 mM $K_3Fe(CN)_6$, 2 mM $K_4Fe(CN)_6$, 0.1% Triton X-100, and 0.04% X-Gluc): 28.80 mL 1 M Na_2HPO_4, 21.20 mL 1 M NaH_2PO_4, 10 mL 0.5 M Na_2EDTA, 20 mL 100 mM $K_3Fe(CN)_6$, 20 mL 100 mM $K_4Fe(CN)_6$, 1 mL Triton X-100, and 0.40 g X-Gluc (*see* **Note 1**) to complete volume to 1000 mL.
7. Chloral hydrate solution: 200 g chloral hydrate, 20 g glycerol, and 50 mL H_2O (*see* **Note 2**).

2.4 Solutions for ClearSee

ClearSee solutions are modified from the previous protocol [12].

1. Phosphate-buffered saline (PBS) solution (10×) (1.3 M NaCl, 70 mM Na_2HPO_4, 30 mM NaH_2PO_4, pH 7.0): 70 mL 1 M Na_2HPO_4, 30 mL 1 M NaH_2PO_4, and 75.97 g NaCl to complete volume to 1000 mL.
2. PBS solution (1×): 1/10 dilution of 10× PBS.
3. 4% Paraformaldehyde (PFA): add 8 g PFA powder to 180 mL 1× PBS solution (*see* **Note 3**). Cool solution to room temperature (RT), and add 1× PBS to complete volume to 200 mL.

Adjust pH to 7.2–7.4 by dropwise addition of 1 M HCl. Store at 4 °C for no longer than 1 month.

4. ClearSee reagent: 10% xylitol, 15% sodium deoxycholate, and 25% urea. To prepare 100 mL solution, add 10 g xylitol powder, 15 g sodium deoxycholate, and 25 g urea to 30 mL H_2O while stirring continuously. Add H_2O to complete volume to 100 mL. Store at RT in darkness.
5. 0.5% Congo red solution: dissolve 0.05 g Congo red solution in 10 mL ClearSee reagent. Store at RT in darkness.

2.5 Microscopy

1. For DIC observations: Nikon Eclipse Ti Microscope (Nikon, Tokyo, Japan).
2. For GFP observations: Zeiss LSM880 confocal laser scanning microscope (Zeiss, Wetzlar, Germany).

3 Methods

3.1 Plant Cultivation (See Note 4)

1. Surface-sterilize *Arabidopsis* seeds with 75% ethanol for 15 min, and then wash with sterilized H_2O four times, each for 20 min.
2. Stratify surface-sterilized seeds at 4 °C for 2 days to ensure simultaneous germination.
3. Transfer the seeds to Petri dishes containing ½ MS medium, and then place the dishes in a plant growth chamber (Percival, Perry, GA, USA) at 22 °C under a 16-h light (~5000 Lux, cool white fluorescent lamp)/8-h dark photoperiod.

3.2 Regeneration Assay

1. Sow *A. thaliana* seedlings on ½ MS medium in a square dish. Suspend the seeds gently in H_2O, and then draw the liquid containing the seeds into a pipette and release one seed at a time, ensuring that they are evenly distributed on the surface of the medium.
2. After 12 days of incubation (22 °C, 16-h light/8-h dark), cut the first pair of rosette leaves from the 12-d-old seedlings at the position between the petiole and the blade (*see* **Note 5**).
3. Culture the detached blades on B5 medium in darkness or B5 medium without sucrose under light conditions. Ensure that the entire detached leaf explant is exposed to the medium (*see* **Note 6**).
4. Adventitious roots emerging from the midvein of the leaf explants near the wound site will be clearly visible by 6–8 days of culture [7].

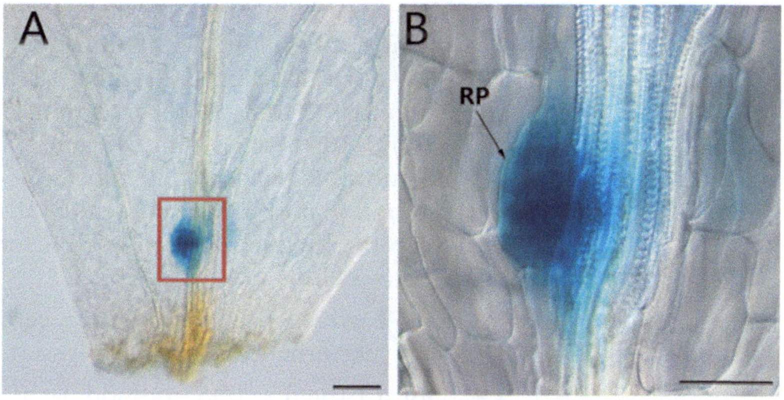

Fig. 2 GUS staining during DNRR from the leaf explant of *WOX5*$_{pro}$*:GUS*, visualized by DIC microscopy. (**a**) Leaf explant of *WOX5*$_{pro}$*:GUS* was cultured on B5 medium in darkness for 4 d. (**b**) Magnifications of boxed region in (**a**). RP, root primordium. Scale bars, 100 μm in (**a**) and 200 μm in (**b**)

3.3 Observation of Gene Expression Patterns by GUS Staining

The expression patterns of key genes in the DNRR process can be analyzed by tracing GUS-labeled genes by DIC microscopy [1].

1. Fix the leaf explants in 80% acetone for 15 min (*see* **Note** 7), and then wash twice with GUS staining buffer.
2. Transfer the leaf explants into the GUS assay solution, and incubate at 37 °C for 0.5–6 h until the blue stain becomes visible (*see* **Note 8**).
3. Remove the stain solution and add 75% EtOH. Change 75% EtOH twice over a 24-hour period until chlorophyll is removed from leaf explants.
4. Transfer the stained tissues into chloral hydrate solution, and incubate at 65 °C for approximately 12 h.
5. When tissues become transparent, trace GUS signals by observation under a DIC microscope (*see* **Note 9**) (Fig. 2).

3.4 Observation of Gene Expression Patterns by ClearSee–Congo Red Staining Method

Here, we introduce a modified ClearSee method to observe gene expression patterns during the DNRR process. The original ClearSee technique has been used for deep imaging of fluorescent proteins in plant tissues and can diminish chlorophyll autofluorescence while maintaining fluorescent protein stability [12]. We improved this method by adding a Congo red staining step to specifically mark the cell wall. This modified ClearSee method allows for better observation of gene expression patterns during the regeneration process by live imaging of proteins labeled with fluorescent tags.

3.4.1 Tissue Fixation and Clearing

1. Add about 15 mL 4% PFA to a 50-mL Falcon tube. To ensure full immersion, wrap the detached leaves in two pieces of filter paper, and fully immerse the wrapped leaves in 4% PFA (*see* **Note 10**).

2. Vacuum-infiltrate wrapped leaves with PFA at RT in a stepwise manner, as follows: 800 mbar, 1 min; 600 mbar, 1 min; 400 mbar, 1 min; 200 mbar, 1 min; 150 mbar, 2 min; 100 mbar, 2 min; 80 mbar, 2 min; 60 mbar, 2 min; 50 mbar, 3 min; 40 mbar, 3 min; 35 mbar, 3 min; 30 mbar, 4 min; 25 mbar, 4 min; and 20 mbar, 20 min. After this vacuum-infiltration procedure, release the valve and let air in very slowly. Keep the leaves immersed in PFA for 1.5 h at room temperature (*see* **Note 11**).
3. Add 1 mL 1× PBS into a 2-mL centrifuge tube. Transfer the leaves into the PBS using forceps.
4. Discard the 1× PBS and add 1 mL ClearSee reagent to the tube. Mix by gently inverting the tube.
5. Incubate at RT in the dark for at least 1 week (*see* **Note 12**). The leaves will slowly become transparent. Once totally transparent, the leaves can be stained with Congo red.

3.4.2 Congo Red Staining and CLSM

1. Transfer 5–10 transparent leaves into 500 μL 0.5% (w/v) Congo red solution, and stain for 2 h to overnight at room temperature.
2. Discard the staining solution, and add another 500 μL ClearSee reagent to destain for 30 min.
3. Place 2–5 leaves on a microscope slide, add drops of ClearSee reagent, and cover with a cover slip.
4. After staining, eGFP fluorescence and the Congo red-stained cell wall can be visualized by CLSM (Fig. 3).

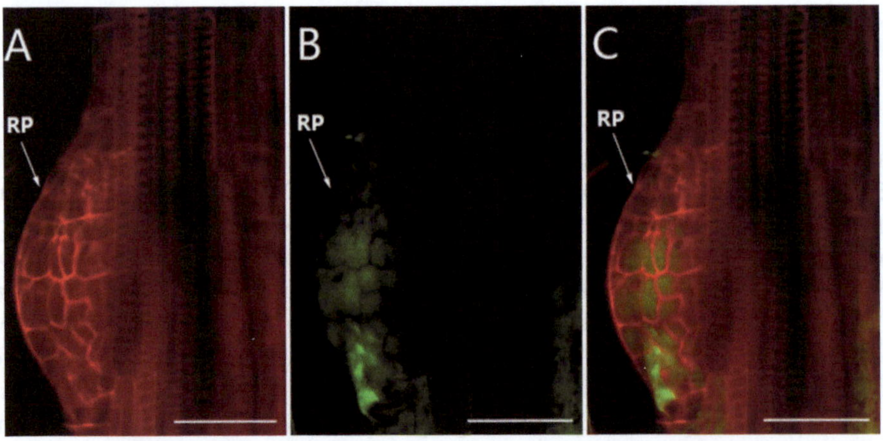

Fig. 3 Analysis of eGFP fluorescence during root regeneration process from leaf explant of *WOX5*$_{pro}$*:eGFP*. (**a–c**) Leaf explant of *WOX5*$_{pro}$*:eGFP* was cultured on B5 medium in darkness for 4 d. Fluorescence was observed in the root primordium by ClearSee–Congo red staining, showing Congo red staining (**a**), eGFP fluorescence (**b**), and merged image (**c**). RP, root primordium. Scale bars = 50 μm

4 Notes

1. X-Gluc is light sensitive. Dissolve powder in DMSO and store at −20 °C in the dark.
2. Mix with a magnetic stirrer for 2–4 h to ensure even mixing of all components, and then store at 4 °C.
3. Prepare 4% paraformaldehyde solution freshly just before use. The PFA will start to dissolve slowly, typically over a period of 30–90 min. Heat mixture at 55 °C, stir every 20–30 min, and add 1 M NaOH dropwise while stirring continuously until all PFA is dissolved. Prepare the solution in a fume hood.
4. For further analyses of root regeneration from leaf explants, it is essential to keep all materials for plant culture under strictly aseptic conditions. All media and materials should be sterilized before use.
5. It is important to cut all the leaves at the same point between the petiole and blade to ensure that adventitious roots regenerate simultaneously.
6. Ensure that leaf explants are sufficiently exposed to the medium to prevent dehydration, and use a gas permeable membrane or tape instead of a parafilm membrane to seal the petri dish. Also, a carbohydrate energy source (e.g., sucrose) must be provided for rooting induction in the dark, because the regeneration process consumes energy
7. In the GUS fixation process, leaf explants should be placed under vacuum for 20 min. Increase and release the pressure slowly. If the samples do not sink to the bottom, repeat the vacuum-infiltration procedure.
8. The GUS staining time depends on the gene expression level. To avoid overstaining by excessive incubation time, check the staining level of the sample regularly.
9. To further trace the cellular localization of GUS signals during the DNRR process, embed samples in resin, cut with a microtome, and observe cut sections by DIC microscopy.
10. Gently shake the glass bottle from time to time to ensure that the samples are immersed in the PFA solution. This will improve the fixation efficiency.
11. After the vacuum-infiltration step, the leaves will sink to the bottom of the tube. If the samples do not sink to the bottom, repeat the vacuum step.
12. The solution will gradually turn green. Change the ClearSee reagent once more if necessary.

Acknowledgments

We thank B. Scheres for *Arabidopsis* lines. This work was supported by grants from the Strategic Priority Research Program "Molecular Mechanism of Plant Growth and Development" of CAS (XDPB0403) and China Postdoctoral Science Foundation (2017M611627).

References

1. Liu J et al (2014) WOX11 and 12 are involved in the first-step cell fate transition during de novo root organogenesis in Arabidopsis. Plant Cell 26:1081–1093
2. Xu L, Huang H (2014) Genetic and epigenetic controls of plant regeneration. Curr Top Dev Biol 108:1–33
3. Ikeuchi M, Ogawa Y, Iwase A, Sugimoto K (2016) Plant regeneration: cellular origins and molecular mechanisms. Development 143:1442–1451
4. Kareem A et al (2016) De novo assembly of plant body plan: a step ahead of Deadpool. Regeneration 3:182–197
5. Chen L et al (2016) YUCCA-mediated auxin biogenesis is required for cell fate transition occurring during de novo root organogenesis in Arabidopsis. J Exp Bot 67:4273–4284
6. Chen X et al (2016) Auxin-independent NAC pathway acts in response to explant-specific wounding and promotes root tip emergence during de novo root organogenesis in arabidopsis. Plant Physiol 170:2136–2145
7. Chen X et al (2014) A simple method suitable to study de novo root organogenesis. Front Plant Sci 5:208
8. Hu X, Xu L (2016) Transcription factors WOX11/12 directly activate WOX5/7 to promote root primordia initiation and organogenesis. Plant Physiol 172:2363–2373
9. Yu J, Liu W, Liu J, Qin P, Xu L (2017) Auxin control of root organogenesis from callus in tissue culture. Front Plant Sci 8:1385
10. Haseloff J, Siemering KR, Prasher DC, Hodge S (1997) Removal of a cryptic intron and subcellular localization of green fluorescent protein are required to mark transgenic Arabidopsis plants brightly. Proc Natl Acad Sci U S A 94:2122–2127
11. Kim MK et al (2002) Specimen block counterstaining for localization of GUS expression in transgenic arabidopsis and tobacco. Plant Cell Rep 21:35–39
12. Kurihara D, Mizuta Y, Sato Y, Higashiyama T (2015) ClearSee: a rapid optical clearing reagent for whole-plant fluorescence imaging. Development 142:4168–4179
13. Muller SM, Galliardt H, Schneider J, Barisas BG, Seidel T (2013) Quantification of Forster resonance energy transfer by monitoring sensitized emission in living plant cells. Front Plant Sci 4:413
14. Truernit E et al (2008) High-resolution whole-mount imaging of three-dimensional tissue organization and gene expression enables the study of phloem development and structure in Arabidopsis. Plant Cell 20:1494–1503
15. Haseloff J (2003) Old botanical techniques for new microscopes. BioTechniques 34:1174–1178, 1180, 1182
16. Kumar R, Silva L (1973) Light ray tracing through a leaf cross section. Appl Opt 12:2950–2954
17. He C, Chen X, Huang H, Xu L (2012) Reprogramming of H3K27me3 is critical for acquisition of pluripotency from cultured Arabidopsis tissues. PLoS Genet 8:e1002911
18. Bennett T, van den Toorn A, Willemsen V, Scheres B (2014) Precise control of plant stem cell activity through parallel regulatory inputs. Development 141:4055–4064
19. Murashige T, Skoog F (1962) A revised medium for rapid growth and bio assays with tobacco tissue cultures. Physiol Plant 15:473–497
20. Gamborg OL, Miller RA, Ojima K (1968) Nutrient requirements of suspension cultures of soybean root cells. Exp Cell Res 50:151–158
21. Tsuge T, Tsukaya H, Uchimiya H (1996) Two independent and polarized processes of cell elongation regulate leaf blade expansion in Arabidopsis thaliana (L.) Heynh. Development 122:1589–1600

Chapter 5

Labeling and Sorting of Arabidopsis SAM Cell Populations to Capture Their Transcriptome Profile

Monika Mahajan and Ram Kishor Yadav

Abstract

In higher plants, the cells that form aboveground tissues and organs are derived from the shoot apical meristem (SAM). SAM is dynamic in nature and divided into central zone (CZ), peripheral zone (PZ), and rib meristem (RM). Stem cells reside in the CZ, and their progenitors differentiate to form lateral organs in PZ and stem tissue in RM. Besides zones, the SAM is also divided into distinct clonal cell layers that show patterned cell division. Here, we describe methods to tag and isolate cell types from both cell layers and zones of SAM in high purity using cell sorter. This method enable plant biologist in rapid isolation of desired cell types from SAM to record their transcriptome, epigenome, proteome, and metabolome. The information generated by this approach will elucidate the mechanism of stem cell self-renewal, differentiation, and organogenesis in SAM.

Key words GFP, Stem cells, FACS, Sorting, SAM, Cell types

1 Introduction

Higher plants maintain continuous growth at both shoot and root ends by the presence of meristematic cells. SAM harbors pluripotent stem cells in CZ, which self-renew and give rise to other cell types that upon differentiation form lateral organs in PZ. To understand the mechanisms of cell division and differentiation in SAM, it is imminent that we develop methods that enable us in labeling the cell population of choice from CZ, PZ, and RM as well as from cell layers [1, 2]. Through cell sorting technique, one can apply to collect cells in adequate quantity to examine spatial and temporal regulation of gene expression at single cell type resolution.

One of the key challenges in multicellular organism is how to label a cell type of interest. This problem is partly solved by the improvement in fluorescent protein folding, solubility, and innovation in transgenic technology. The green/red fluorescent protein is expressed under a cell- or tissue-specific promoter, and these proteins localize to the nucleus or endoplasmic reticulum depending

Muhammad Naseem and Thomas Dandekar (eds.), *Plant Stem Cells: Methods and Protocols*, Methods in Molecular Biology, vol. 2094, https://doi.org/10.1007/978-1-0716-0183-9_5,

© Springer Science+Business Media, LLC, part of Springer Nature 2020

upon the attached signal peptide to prevent their free diffusion. In model plant *Arabidopsis thaliana*, several well-characterized promoters are reported in literature, and that were used to tag cell types in specific tissues of adult plant. In many studies, the above approaches have been used successfully to generate fluorescent protein-based stable transgenic plant lines [3].

Once the transgenic plant lines are established, they are grown for cell sorting purposes, and tissue is used for isolation of desired cells after subjecting SAM to protoplasting treatment (*see* **Note 1**). The protoplasting method developed for root tissue did not work initially for SAM tissue owing to the differences in the complexity of cell wall composition between the shoot and root. To overcome this problem, the concentration of pectolyase was increased, and moreover, in addition to cellulase, hemicellulase was added in the protoplasting solution [4].

In this chapter, we describe detailed description of the methods that were developed for marking cells in SAM with fluorescent proteins for their isolation in high purity using fluorescent-activated cell sorter (FACS). This chapter covers both making of transgenic plant lines and their screening, as well as cell sorting. The description on RNAseq is insignificant, given the fact that this technique is evolving continuously, and a variety of platform and methods of library preparation are at disposal to an experimenter. Therefore, in future sequencing of extremely low quantities of RNA will be possible with advancement of technology.

2 Materials

1. **Fluorescent reporter plant lines:** In model plant Arabidopsis, fluorescent protein-based reporters are generated by creating stable transgenic lines routinely. For this purpose, one can choose a promoter of a gene whose expression is confined to specific cell or tissue type. For shoot apex cell types, well-characterized transcripts have been reported in the literature; promoter of such genes can easily be adapted to make fluorescent reporter lines. A variety of fluorescent proteins are available, and in principle, one can generate fluorescent reporter using any of them (GFP, YFP, Ypet, RFP, Td-tomato, etc.). A tag is attached to fluorescent protein either at the N-terminus or C-terminus (such as H2B, NLS, ER); this will prevent their diffusion to the neighboring cells. Thus, it will ensure further purity of the sorted cell population.
2. **Confocal microscope with water dipping lens:** To identify the right transgenic line for cell sorting, T1 transformants are subjected to shoot imaging using an upright confocal microscope equipped with water dipping lens. In our study, we have

used *apetala 1-1;cauliflower 1-1* (*ap1-1;cal1-1*) double mutant; therefore, there is no need for dissection of the shoot before imaging. The water dipping lens with long work distance is ideal to screen the fluorescent protein reporter lines. Objectives having 40, and 63× magnifications with numerical apertures ranging from 0.9 to 1.1, are excellent for high-quality imaging of shoot apices.

3. **Fluorescent-activated cell sorter (FACS):** FACS equipped with appropriate excitation laser (488, 561, 631 nm) combination is required for sorting. Two-way cell sorter is sufficient to carry out most routine sorting at high purity. Most of the available cell sorters nowadays are equipped with 3–4 lasers and can sort cells in two, three, and four ways depending upon the requirement of user.
4. **Refrigerated shaker:** To ensure the proper temperature to tissues subjected to protoplasting for cell sorting, orbital motion shaker is required. Alternatively, plant growth chambers equipped with electric plug can also be made to work by putting a small benchtop orbital shaker.
5. **Refrigerated centrifuge:** Refrigerated centrifuge for pelleting down the cells after protoplasting treatment.
6. **Tweezers:** Fine tweezers for harvesting the shoot apices (e.g., Dumont 5 INOX).
7. **Sterile filters**: 40 μm and 70 μm cell strainer mesh.
8. **Sterile tubes**: 15 mL and 50 mL capacity falcon tubes and 5 mL polystyrene round-bottom tubes with a 35 μm cell strainer snap cap.
9. **Ethanol**: 70% ethanol and absolute ethanol.
10. **Eppendorf tubes**: RNase- and DNase-free eppendorf tubes and DNA LoBind tube for storage and transport of RNA sample.
11. **Reagents**: RNase-free water, Plant RNeasy Mini Kit (QIAGEN), DNase treatment kit (QIAGEN), GlycoBlue (Invitrogen), 7.5 M ammonium acetate, 1 M KCl, 1 M $MgCl_2$, 1 M $CaCl_2$, 1 M Tris without pH adjustment, bovine serum albumin, 2-(**N**-morpholino)-ethanesulfonic (MES) acid, mannitol, cellulase (Yakult), macerozyme (Yakult), and hemicellulase (Sigma). **Solution A:** Prepare fresh solution A in 50 mL falcon, for 25 mL, add 250 μL 1 M KCl, 50 μL 1 M $MgCl_2$, 50 μL 1 M $CaCl_2$, 0.025gm BSA, 0.0097gm MES, and 2.75gm mannitol. Add 20 mL dH_2O, and adjust pH to 5.5 using 1 M Tris (unadjusted) before making final volume to 25 mL. **Solution B**: Weigh 300 mg cellulase, 200 mg macerozyme, and 200 mg hemicellulase in a 50 mL falcon, and add 20 mL solution A in it. Dissolve by mixing gently the content of the falcon tube at

ambient temperature. Heat the solution to 55 °C for 10 min (solution will turn clear), and cool it down to room temperature before adding 0.1% w/v BSA, 10 mM $CaCl_2$, and 5 mM β-mercaptoethanol. Store at 4 °C.

3 Methods

1. *The ap1–1;cal1–1* is an ideal mutant for isolating distinct cell types including stem cells from the shoot apical meristem. Plants carrying the transgene for promoter-reporter fusion are harvested on the aluminum foil quickly to avoid spurious effect. Tissues weighing around 50–60 mg are sufficient to sort close to 100,000 cells.
2. Plants are grown in controlled environment condition at 22–23 °C for about 4 weeks in soilrite mix supplemented with compost and nutrients. A sorting experiment requires ~200 shoots, which is typically one tray of plants. When the plants start forming cauliflower-like heads and still appears green, it is the right time to begin with the sorting experiment. Harvest the shoot apices with the help of tweezers gently, and place them on the aluminum foil. It takes less than 5 min to collect a single tray of plants.
3. Harvested apices are transferred into a 50 mL falcon tube containing 7–8 mL protoplasting cocktail (solution B). The SAM tissue will take time to sink in the solution; at this point, gentle mixing by hand rotating the tube upside down would assist in the penetration of solution in the tissue (*see* **Note 2**). Use a benchtop orbital shaker kept at 22 °C or orbital shaker equipped with temperature control. The speed of the shaker needs to be adjusted between 110 and 120 rpm.
4. Keep the contents shaking for an hour. Gently make the tube upside down several times, at interval of 10–15 min; this will facilitate the mixing of protoplasting enzymes thoroughly in the plant tissue.
5. Take a sterile petri dish, and place 40 μm cell strainer in it and pour the contents of the falcon tube. Remove the cell strainer along with undigested SAM tissue. Transfer the protoplast suspension from petri dish to 15 mL falcon tube with a wide-mouth pipet tip. Keep the cell suspension on ice.
6. Pellet down the protoplasts by centrifugation at 500 × *g* for 10 min at 4 °C. Decant the supernatant carefully without disturbing the pellet. To wash the cell pellet, add 1 mL of solution A, resuspend cells, and repeat the above step. Depending on the size of pellet, add 500–1000 μL of solution A (*see* **Note 3**), cut the pipette tip, and gently dissolve the protoplast

without leaving any clumps. Transfer the contents of the tube into a 1.5 mL sterile eppendorf tube, and keep it on ice.

7. A few microliters of cell suspension can be put on a hemocytometer to check the density and viability of protoplasts. In order to ensure high cell yield and purity, cell density is adjusted by diluting the cell suspension with solution A. The cell density in the range of ~10^6 cells/mL is ideal for sorting (*see* **Note 3**). Finally, cell suspension is passed through snap cap of 5 mL polystyrene round-bottom tube fitted with nylon mesh. Once the sample is filtered, it is ready for sorting.
8. Prepare always a negative control consisting of nonfluorescent cells in parallel to measure forward (FSC) and side scatter (SSC) of the given cell population. Next, dot plot for protoplast isolated from fluorescently labeled plants (*pCLV3::mGFP-ER, ap1–1;cal1–1*) is prepared, and cell population of fluorescent-labeled protoplasts is selected by the emission spectra of the fluorochrome in their respective channels above the negative control.
9. A protoplast suspension derived from fluorescent marker line will produce a clear population of fluorescent events, which are not observed in nonfluorescent sample. Perform doublet exclusion by plotting the width versus area for forward and side scatter. Finally, apply the gate to identify the fluorescent cells by comparing with the control nonfluorescent cells, and start sorting (Fig. 1c and d).
10. Sorting is usually performed with a 100 μm nozzle at the rate of 2000–5000 events per second under sheath pressure of 20 psi. The purity and integrity of sorted cells can be verified again by putting a few microliters of sorted cells in epifluorescence microscope (Fig. 1e, f, g, and h).
11. To isolate RNA, cells positive for fluorescence are sorted directly into 500 μL RLT buffer. When the number of cells reaches close to ~100,000, the eppendorf tube can be replaced with another fresh tube containing 500 μL RLT buffer (optional, add 5 μL of β-ME and store cells in −80 °C).
12. To isolate RNA, adjust the volume of cell suspension to 3 mL by adding RLT buffer in a 15 mL falcon tube. Add 30 μL of β-ME and 0.5 volume of 100% chilled ethanol in the tube. Mix the contents gently up and down and keep on ice. RNA will precipitate after adding the ethanol, and the contents of the tube will be passed through RNA binding column.
13. Take a RNeasy column and add ~700 μL of precipitate, centrifuge at 10,621 × *g* for 10 s at 4 °C, and repeat the same step again until the whole precipitate passes through the column. The RNeasy columns withstand the repeated centrifugation

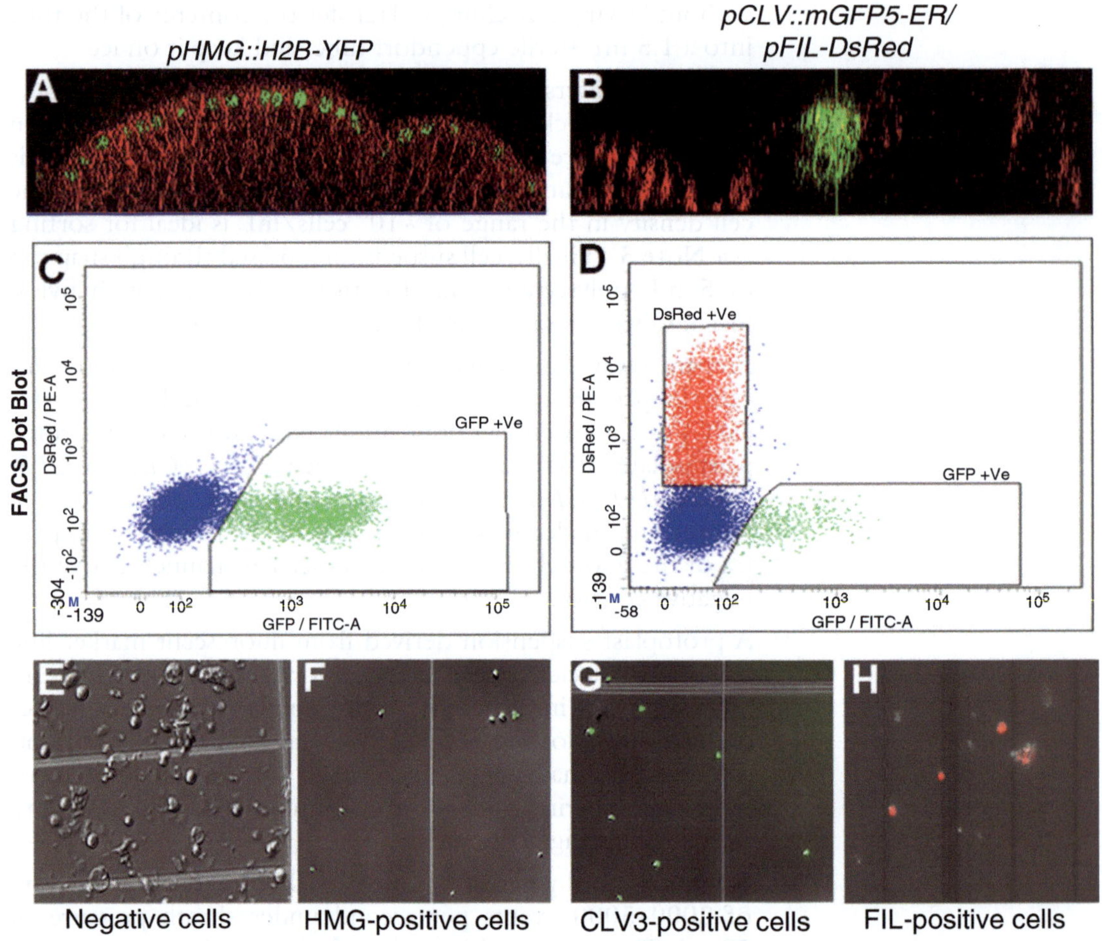

Fig. 1 Labeling and sorting of cells for gene expression studies. A side view of SAM showing the expression of *pHMG::H2B-YFP* in L1 layer (**a**). *pCLV::mGFP5-ER* and *pFIL::DsRed* in central and peripheral zone, respectively (**b**). The FACS-generated scatter plot shows a population under the excitation parameter (488 nm) in which a proportion of cells are GFP positive (green, **c**), and the scatter plot shows a population under the excitation parameters (488 nm, 561 nm) in which a proportion of cells are GFP positive (green), Ds-Red positive (red), and negative cells (blue). (**d**) Epifluorescence microscopy image of FACS-sorted cells from *pHMG::H2B-YFP* (**f**), *pCLV3::mGFP5-ER* (**g**), and *pFIL::DsRed* (**h**) reporter lines. GFP-ve cells are shown in (**e**). Images were obtained in fluorescence and DIC mode

force, but care must be taken to ensure the integrity of the RNA-binding resin.

14. To wash the RNeasy column, apply 350 μL RW1 buffer and centrifuge it for 15 s at 10,621 × *g*.
15. On column, DNA digestion is performed by applying DNase in RDD buffer (10 μL DNase + 70 μL RDD buffer). Mix the DNase solution in RDD buffer in a separate tube, and apply in the center of the column; leave it at room temperature for 15 min. To proceed further, wash the column again with 350 μL RW1 buffer, and spin the column at 10,621 × *g* for 10 s.

16. Add 500 μL RPE buffer to column, and centrifuge it for 15 s at 10,621 × *g*. Decant the flow-through and spin once more for 2 min at 10,621 × *g* to dry the column.
17. Transfer the column to an RNase-free sterile eppendorf, add 25 μL of RNase-free water to column, and spin it at 17,949 × *g* for 1 min. Repeat the same step once again.
18. The concentration of total RNA/μL will be very poor. To concentrate RNA, precipitation is required. Add 0.5 volume of 7.5 M ammonium acetate, 2.5 volume absolute ethanol (100%), and 0.5 μL GlycoBlue (15 mg/mL) (*see* **Note 5**), and after mixing the content of the tube, leave it in −20 °C for overnight.
19. The next morning, pellet down the RNA for 30 min at 4 °C and 17,949 × *g*. Remove the supernatant without leaving any trace of it in the tube. Add 100 μL of 70% ethanol in the tube, and spin it 17,949 × *g* for 5 min at 4 °C. Remove the supernatant by long tip gently, and let the tube get dry for 5–10 min at room temperature. Add 5–10 μL of RNase-free water to dissolve RNA, and store it in −80 °C.
20. RNAseq: Nowadays, RNAseq has become affordable to sequence low quantities of RNA. A variety of service providers based on the latest platform of Illumina are providing the services to the community. Most service providers are capable of sequencing RNA as pair-end reads, which will be useful in identifying novel transcripts and alternative splicing variant in cell types of interest.

4 Notes

1. To reproduce the high-quality transcriptome data for individual biological replicates, care must be taken from the very beginning, e.g., growing plants, protoplasting, cell sorting, etc. Make sure the plant line used for cell sorting descended from a single parent, and expression pattern of the reporter is close to native gene expression pattern. Seeds bulked up from single parent have to be used for all biological replicates. A few representative plants can be randomly checked for assessment of reporter expression and integrity. Plants can be grown in standard Arabidopsis growth chambers as well as in plant growth room designed for Arabidopsis growth. Plants need to be grown at equal distance; per pot, do not put more than 8–10 plantlets. We grew plants for 4 weeks.
2. Transcript levels may get influenced by the time of the day. Therefore, all cell sorting experiments should be done preferably at the same time of the day. Many factors can influence the

cell population-specific transcriptome; therefore, to minimize lot-to-lot variation, time taken for harvesting tissue should be the same, and temperature of the room where plants shoot apices are harvested needs to be set at 22 °C. Sixty minutes of protoplasting treatment is applied in most cases. However, often, tissue does not sink properly in the protoplasting cocktail; this would result in poor digestion of cell wall, and eventually, the efficiency of protoplasting goes down. Thus, the total number of cells will be very less. One can easily check the turbidity of the solution inside the falcon tube to assess the progress of protoplast release. Mix tissue in the solution B thoroughly in the very beginning, gently by turning the tube up and down several times. Check every 10–15 min progress of protoplasting looking at the turbidity of solution, and slowly you would notice that the solution has started turning turbid due to release of cells.

3. Too much high density of the protoplast in the loading tube can lead to the clogging of FACS. Make sure the sample is diluted in appropriate range of cell density before loading.
4. Doublet discrimination is important for sorting the fluorescent cell population. Due to poor digestion of cell wall, a fluorescence-positive cell stays with negative cell. When such a doublet made up of two cells encounters the laser, it will excite the fluorochrome and will sort the doublet in the positive sample. Doublets are distinguishable when compared with singlets based on their width and area.
5. RNeasy isolation kit provides total RNA yield in the range of 200–300 ng. Incomplete precipitation or fewer number of cells sometimes may result in poor yield of total RNA. Make sure the contents of the tubes are mixed thoroughly after adding ethanol. GlycoBlue adds in visibility of RNA pellet and also increases the overall yield of the precipitated RNA.
6. If the promoter expresses in a few cells, it does affect the overall yield of fluorescently labeled cells in a FACS experiment. In our hands, we managed to do gene expression studies using ~30,000 cells. Alternatively, one can grow more plants in parallel to increase the overall yield of cells.
7. To maintain the high integrity of RNA, care must be taken while drying and dissolving the RNA. Excessive drying of RNA at room temperature or by heating results in insolubility of RNA. Vigorous shaking of partially dissolved RNA or vortexing may result in the disintegration of RNA, which is unsuitable for RNAseq. To avoid this, after pelleting down the RNA, remove the supernatant using long tip without disturbing the pellet. Let the tube dry at room temperature for 5 min. Add RNase-free water immediately, and tap the tube gently at the

bottom; you may keep it in thermostat for 2 min at 60 °C. The RNA pellet dissolved this way will have good yield and high quality. To ensure the optimum recovery later on, use always DNA LoBind tubes.

8. Gene expression of sorted cells does get influenced by protoplasting treatment, sorting time, etc. Therefore, the experimenter should make sure that time taken for each step should be the same across different biological replicate.

References

1. Steeves Taylor A, Sussex IM (1989) Pattern in plant development. Cambridge University Press, New York
2. Satina S, Blakeslee AF, Avery AG (1940) Demonstration of the three germ layers in the shoot apex of datura by means of induced polyploidy in periclinal chimeras on JSTOR. Am J Bot 27(10):895–905. https://doi.org/10.2307/2436558
3. Brady SM, Orlando D, Lee JY, Wang JY, Koch J, Dinneny JR, Mace D, Ohler U, Benfey PN (2007) A high-resolution root spatiotemporal map reveals dominant expression patterns. Science 318:801–806
4. Yadav RK, Girke T, Pasala S, Xie M, Reddy GV (2009) Gene expression map of the Arabidopsis shoot apical meristem stem cell niche. Proc Natl Acad Sci U S A 106:4941–4946

Chapter 6

Plant-Associated Microbes Alter Root Growth by Modulating Root Apical Meristem

Anwar Hussain, Husna, Ihsan Ullah, and Muhammad Naseem

Abstract

Rhizobacteria are known to produce a variety of signal molecules which may modify plant growth by interfering with phytohormone balance. Among the microbial signals are phytohormones, known to contribute to plant endogenous pool of phytohormones. The current chapter describes different methods to study the regulation of gene expression in root apical meristem in response to rhizobacterial inoculation. We describe protocol for the detection of in planta modulation of CKs and IAA by rhizobacteria and their impact on root growth, dissecting the underlying plant signaling pathway by RNA sequencing.

Key words RAM, Cytokinins, Auxin, Rhizobacteria, Root growth

1 Introduction

The plant root is an important organ that supports plant stationary mode of life by serving a great variety of functions, from anchoring in soil to shaping microbial community in the rhizosphere for sustaining favorable immediate environment. The plant root releases root exudates, a treasure of secondary metabolites, used to differentially communicate to friends and foes [1].

The meristem is a developmental plant tissue made up of self-renewing, undifferentiated, and pluripotent stem cells responsible for plant growth and development [2]. Due to their central role in plant life, meristems have been a hot spot of scientific research for more than 150 years. Apical meristem is an important meristem partitioned into the root and shoot apical meristems. The shoot apical meristem (SAM) provides cells of all aerial postembryonic organs, while the root apical meristem (RAM) does the same function in undergrounds promoting root growth [3].

The root apical meristem has a well-defined structure showing radial and longitudinal stereotypical patterns of cell types. The pattern on the radial axis is obvious in concentric rings composed of the lateral root cap, the epidermis, the ground tissue (cortex and

Muhammad Naseem and Thomas Dandekar (eds.), *Plant Stem Cells: Methods and Protocols*, Methods in Molecular Biology, vol. 2094, https://doi.org/10.1007/978-1-0716-0183-9_6, © Springer Science+Business Media, LLC, part of Springer Nature 2020

endodermis), and a pericycle around a central stele [4, 5]. This pattern is designed as embryogenesis progress and maintained by the activities of four sets of initials produced by the stem cell: the lateral root cap/epidermal, the cortical/endodermal, the columella, and the pericycle/vascular initials surrounding quiescent center (QC) [6, 7]. Root stem cells are capable of prolonged self-renewal and simultaneously providing cells for differentiation [8]. The QC influence one of the two daughters formed as a result of asymmetrical division of the stem cell, to remain meristematic by short-range signals. Contrary to this, the other daughter cell is contributed to the differentiated tissues [9].

The root meristem forms an apical zone, basal transition zone, elongation zone, and maturation zone along the longitudinal axis. The apical zone is the root apical meristem (RAM) composed of the stem cell niche, columellate, lateral root cap, and proximal meristem having actively dividing cells [10]. Due to a stereotyped division pattern in the stem cells, stacks of cells are formed where the spatial arrangement of cells in a column shows their age: older cells are at the top and younger cells at the bottom near the root tip [11].

The root meristem size and growth rate are determined by an intricate balance between the maintenance of stem cells and the degree of cell differentiation. The mechanisms involve in maintaining this balance is only partially understood. According to our current understanding, the organization and maintenance of RAM are controlled by intercellular communication [12].

Rhizobacteria are known to interact with the plant root in the rhizosphere and significantly modulate root growth responses [13, 14]. Microbes can act as a source of phytohormones modulating RAM and subsequently root growth [15]. Besides, these rhizobacteria have the ability to induce changes in the expression of key genes by inducing host to produce novel miRNAs. Thus, by inducing host plant to produce miRNAs, the microbes alter their root growth responses through modulation of phytohormones signaling and the behavior of RAM. The current chapter describes methodology of inoculating compatible plant growth-promoting rhizobacteria and preparation of slides for observing root sections and microdissection of different components of RAM for downstream applications such as RNA sequencing to study gene expression.

2 Materials

2.1 Biologicals

1. *Arabidopsis thaliana* col-0.
2. *A. thaliana* ARR5::GUS.
3. *A. thaliana* DR5::GUS.

4. *A. thaliana* SHY2::GUS.
5. *Pseudomonas aeruginosa* Am3 and *P. aeruginosa* Am2 or any other plant growth-promoting rhizobacteria.

2.2 Chemicals

1. Murashige and Skoog basal salt (Sigma-Aldrich) solidified with 1% agar and containing 3% sucrose.
2. To prepare Luria-Bertani (LB) media, take 400 mL distilled water in 1 L beaker placed having a magnetic bar on a magnetic stirrer set on stirring mode. Sequentially add tryptone (5 g), yeast extract (2.5 g), and NaCl (5 g) to the beaker, and make final volume to 500 mL with distilled water. Adjust pH to 7–7.2.
3. To prepare GUS buffer, add 1 mM 5-bromo-4-chloro-3-indolyl-β-d-glucuronide at pH 7.0 (X-Gluc; Molecular Probes, Eugene, OR, USA) in 50 μL dimethylformamide, and mix well.
4. Dissolve diaminobenzidine tetrachloride in distilled water to prepare a working solution of 2.5 mmol L^{-1} for DAB assay. Prepare a fresh solution just before use.
5. 70% Ethanol.
6. 50% Bleach.
7. 0.1% Ruthenium red.
8. Xylene.
9. Acetone.
10. Paraplast chips.
11. Histoclear (Sigma-Aldrich, USA).
12. Mineral oil.

2.3 Instruments

1. Plant growth chamber.
2. Laser scanning microdissection microscope (LSM).
3. Microtome/cryostat.
4. Light microscope.
5. Shaking incubator.

3 Methods

3.1 Selection of Rhizobacteria

1. To select suitable rhizobacteria, screen the isolates for their effect on plant root. Surface sterilize *Arabidopsis thaliana* col-0 seeds by suspending the seeds in 70% ethanol for 2 min in an Eppendorf tube. Now, suspend the seeds in 50% bleach for 5 min with occasional shaking to resuspend the seeds settled down in the tube. Discard the bleach, and wash the seeds thoroughly with autoclaved distilled water thrice.

2. Put the seeds in Petri plates containing Murashige and Skoog agar (Sigma, Germany) in a line dividing the plate into two asymmetrical halves sparing three-fourth portion of the plate for root growth. Allow the seeds to germinate on vertical plates at 22 °C, 70% relative humidity, and dark. After germination, put bacterial inoculum adjusted to 10^6 cfu/mL^1 directly onto the roots in such a way that each seedling receives 50 μL of inoculum. Allow the seedlings to grow for 1 week at 22 °C, 70% relative humidity, and long day photoperiod (16 h light and 8 h dark cycles) in a growth chamber.
3. Measure root and shoot length of the seedlings by taking pictures of the plates and processing the pictures via ImageJ software (v1.44; US National Institutes of Health, Bethesda, MD).
4. Study root anatomy after extracting the seedlings from the agar plates and washing away the agar. Prepare the root segments (5–10 mm) for section cutting with microtome, and study the RAM for any change.
5. Also, stain the roots from 7-day-old seedlings with 25 mmol L^{-1} diaminobenzidine tetrachloride (DAB; Sigma-Aldrich, St. Louis, MO, USA) by dipping the roots in the stain for 10–12 h. Using aniline blue, counterstain the DAB-stained roots, and observe under the light microscope. DAB staining helps in visualizing the intracellular bacteria as such bacterial cells are surrounded by H_2O_2 [16].
6. Select bacteria able to modulate RAM, and improve root growth.

3.2 Screening Rhizobacteria for In Planta CK and IAA Production

1. Use *Arabidopsis* ARR5::GUS and DR5::GUS lines to screen rhizobacteria for in planta CK and IAA production, respectively.
2. Follow the procedure given in Subheading 3.1, **steps 1** and **2**, to germinate the mutant seeds on agar plates and inoculate rhizobacteria on their roots.
3. Harvest the seedlings after 1 week, and perform GUS staining procedure by putting the root segments in GUS buffer composed of 1 mM 5-bromo-4-chloro-3-indolyl-β-d-glucuronide at pH 7.0 (X-Gluc; Molecular Probes, Eugene, OR, USA) using DMF (50 μL) and 50 mM sodium phosphate buffer pH 7.2 as solvent and diluent, respectively. Add 0.2% Triton X-100 as supplement.
4. Infiltrate the GUS under vacuum for 30 min followed by an incubation at 37 °C for 3–24.

5. Fix the stained root segments with 4–5% glutaraldehyde (v/v) and paraformaldehyde (w/v) in 0.05 M nitrate phosphate buffer (pH 7.2) overnight at 4 °C [17].
6. Dehydrate the fixed root material in an ethanol series, and then embed Technovit 7100 (Heraeus Kulzer) following the manufacturer's instructions.
7. Using microtome, make longitudinal and transverse sections (9–10 μm), and then stain the sections 0.1% ruthenium red.
8. Using a light microscope, analyze the sections for GUS staining in the root segments. Intensively stained areas show bacterial CKs or IAA, respectively, in comparison to control.

3.3 Control of Cell Division and Differentiation in RAM

1. The role of CKs and auxin is indispensable in controlling the meristem activity. Cytokinins activate the gene SHY2/IAA3 (SHY2) that codes for a repressor protein involved in suppressing auxin signaling resulting in negative regulation of the auxin transport facilitator genes of the PIN family. Thus, CKs promote cell differentiation in RAM by auxin redistribution. Contrary to this, auxins sustain PIN activity and promote cell division via degradation of the SHY2 protein. Thus, knowing the expression of SHY2 protein, it is possible to find the balance between cell division and cell differentiation necessary for the RAM activity [18]. Obtain SHY2::GUS lines to show that the impact of rhizobacteria on RAM is due to their ability to produce phytohormones, CKs and IAA.
2. Grow the three sets of *Arabidopsis* mutants, each set including SHY2::GUS and ARR1-3 on vertical plates as described above.
3. Inoculate CK-producing and IAA-producing rhizobacteria on the roots of seedlings in two separate sets, and leave the third set as uninoculated control. Allow the seedlings to grow for 1 week.
4. Perform GUS staining procedure and section cutting as described in Subheading 3.2.

3.4 Molecular Characterization of RAM Organization

1. To study the impact of rhizobacterial inoculation on different elements of RAN, use the laser microdissection techniques for isolation of different components of RAM. To do so, prepare root tissue for the procedure as follows. Cut segments of *Arabidopsis* roots by using RNase-free scalpel, and immediately immerse the obtained segments in a 2 mL RNase-free Eppendorf tube containing ice-cold 100% acetone for fixation. Incubate the samples under vacuum to carry out vacuum infiltration of the fixative for 15 min under 350 mm of Hg pressure at 4 °C. The procedure may be prolonged in case the root segments do not settled down in the solvent at the end of 15 min (*see* **Note 1**).

2. Replace the acetone in the samples by fresh ice-cold acetone, and incubate the tubes at room temperature (RT) for 1 h on a shaker set at 50 rpm. Refresh the acetone once again, and leave the shaking tube overnight at 4 °C.
3. Next morning, prepare a series of dilutions of acetone in xylene, 3:1, 1:1, and 1:3. Dehydrate/infiltrate the root samples by passing the segments through the series of dilutions for 1 h each. Finally, infiltrate the root segments with 100% xylene at room temperature for 1 h.
4. Add 2–3 paraplast chips (Sigma-Aldrich, USA) into each vial, and leave the vials at room temperature for overnight. Now, keep the tubes in oven set at 57 °C in order to melt the paraplast for 3 days. Replace the paraplast in the vials with fresh molten paraplast at regular intervals of 10–12 h (*see* **Note 2**).

3.5 Tissues Embedding in Paraplasts

1. Take Peel-A-Way molds (Sigma-Aldrich, USA), and pour the contents of each vial in a separate mold put on a hot plate at 57 °C, and arrange the root segments with the help of RNase-free forceps in a desired orientation.
2. Prepare small blocks of solidified paraplast with embedded root segments by shifting the molds at room temperature (RT), and then store the blocks at 4 °C [19].

3.6 Root Sectioning

1. For section cutting, trim the root blocks into desired shape, and then place the blocks on plastic embedding ring (Himedia, India) in a desired orientation for cutting longitudinal and transverse sections.
2. Fix the rings with the holding clamp of the rotary microtome (Leica RM2265) to cut 8–10 μm thin root sections (*see* **Note 3**).
3. Flatten the sections put on HistoBond + charged slides (Marienfeld, Germany) on a water bath set at 50 °C for 3 min.
4. Dry the flattened sections at 42 °C for 30 min, and store for LCM use at 4 °C.

3.7 Laser Capture Microdissection for Fractionation of RAM

1. Once thin sections of root segments are made, use histoclear (HistoChoice Clearing Reagent, Sigma-Aldrich, USA) to dewax the sections. To do so, dip the tissue mound slides twice in histoclear for 2 min, and keep the slide at room temperature to air dry the sample LSM (*see* Fig. 1).
2. Observe the prepared slide under light microscope to make sure that the procedure is successfully performed (*see* **Note 4**).
3. Using a laser microdissection microscope (Leica LMD7, Germany), observe the slides put in face down position on the

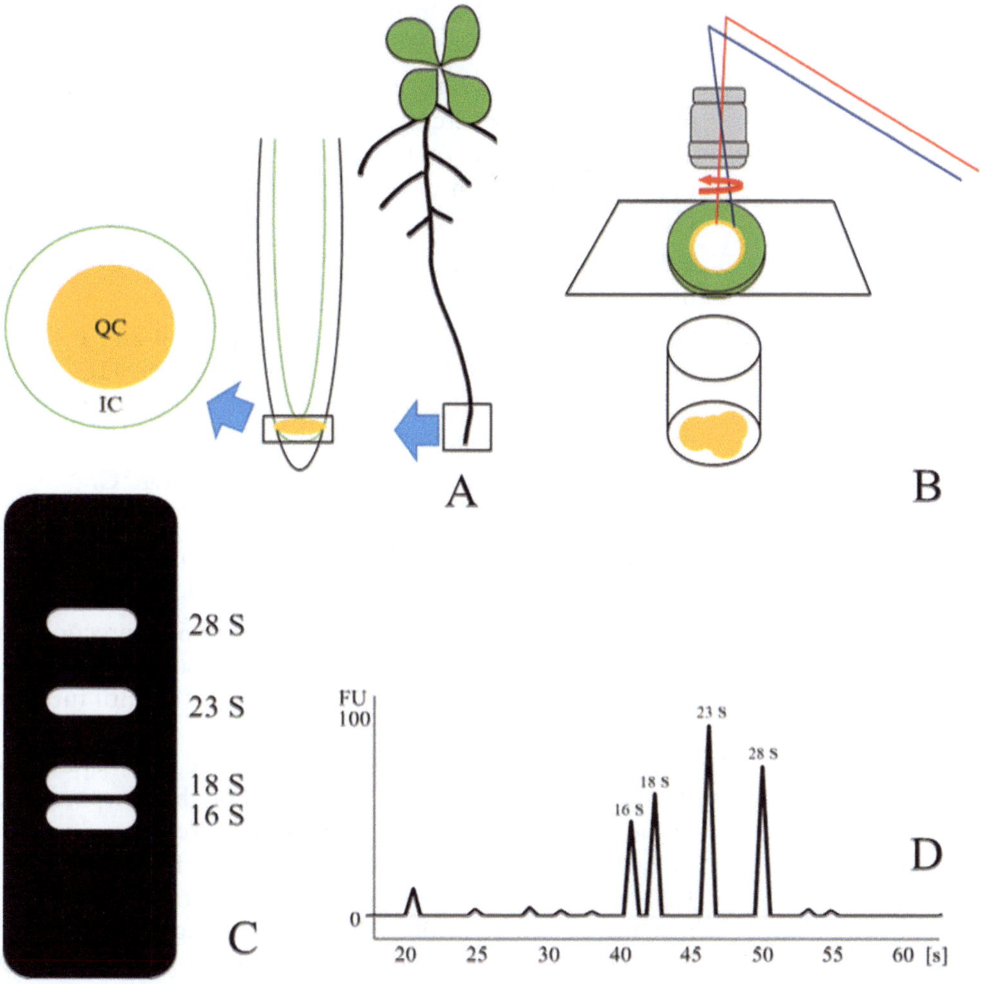

Fig. 1 Some important steps to determine comparative gene expression in the RAM of rhizobacteria-associated *Arabidopsis* seedlings through RNA sequencing. (**a**) Root section through quiescent center (QC) and initial cells (IC), (**b**) obtaining the QC through laser scanning microdissection for RNA isolation, (**c**) Polyacrylamide gel electrophoresis to visualize the isolated, and (**d**) RNA integrity number (RIN) for RNA quality check

microscope, and identify different components of RAM. Mark the different components of RAM using a mouse, and then cut the selected area using UV laser (337 nm) (*see* **Note 5**).

4. Collect the excised tissues in RNase-free 0.5 mL tubes containing a drop of mineral oil (Amresco, USA). Store this sample at −80 °C or subject directly to RNA isolation procedure (*see* **Note 6**).
5. Optimize the dissection conditions for getting clean and narrow excision of the tissues encircled on computer screen [19].

3.8 Isolation of RNA

1. Extract total RNA from different segments of the RAM obtained by LSM (*see* Fig. 1).
2. Before you start with extraction, treat all the plasticwares, pestle and motor, etc. with diethylpyrocarbonate (DPEC) to avoid any RNase contamination. Alternatively, DPEC-treated tips, Eppendorf, and other plasticware may be used (*see* **Note 7**).
3. Take 50 sections from −80 °C, and grind the frozen tissue to fine powder with the help of pestle and motor.
4. Sequentially add 300 μL of ice-cold phenol and 0.1 M TES buffer composed of 1 mM EDTA, 10 mM Tris–HCl, and 0.5% SDS (pH 7.5) to the powdered plant material, and take the mixture in an Eppendorf (*see* **Note 8**).
5. Spun down the pooled aliquots for 5 min at 4 °C.
6. Transfer the top aqueous phase to a fresh Eppendorf leaving the interface undisturbed. Wash the aqueous phase twice with chloroform to remove any leftover phenol.
7. Precipitate the RNA with 2.5 volumes of 59% precool ethanol at −20 °C for 2 h. Resuspend the RNA in approximately 30 μL of DPEC-treated water, and keep at −20 °C until future use.

3.9 RNA Quality Check

1. For getting meaningful gene expression data, evaluation of RNA integrity is a critical first step. For this purpose, the use of bioanalyzer is highly recommended. Prepare a 3 μL aliquot of each RNA sample at a concentration of 50–500 ng/μL in RNase-free H2O. Determine RNA integrity number (RIN) using Agilent RNA 6000 Nano Kit and Agilent 2100 Bioanalyzer (Agilent Technologies, Inc.) following the manufacturer instructions (*see* **Note 9**).
2. Library preparation and RNA sequencing. Subject the RNA to library preparation by using standard protocols provided with the RNA prep kits, and send your samples for RNA sequencing to a company of good repute.

4 Notes

1. Degradation of RNA starts soon after plant tissues are harvested. Hence, it is imperative to carry out vacuum infiltration of the fixative till the tissues settle down at the bottom indicating complete replacement of internal air by the fixative. Due to histological differences and variation in secondary metabolites among different plant tissues, fixation steps may need to be standardized.

2. Prolong incubation at high temperature is another common cause of RNA degradation lowering the quality of isolated RNA. Lowering temperature to just above the melting point (57 °C) improves the quality of RNA.

3. Care should be taken to cut section perpendicular to the longitudinal or transverse axis of the root. It is important because an angled section means that all the succeeding sections will be angled. Such sections and cells within the sections are oval, and some parts of the section show smeared appearance. Another important problem with section cutting is to make too thin sections which have missing parts and/or not completely round. Also, use of dull razor blades should be avoided because such practice results in smeared and misshaped sections. To avoid contamination in RNA, use new cryostat blade for each new root segment.

4. Tissues with a frayed appearance show digestion of the tissues in cases of exposure to strong fixative or prolonged exposure to fixative.

5. Care was taken to direct the laser to ablate the cells surrounding the selected cells, therefore preserving their integrity. Off target firing of laser shows that the objective magnification used for laser focusing is different than magnification used for capture and cell annotation, or location of laser is incorrectly set.

6. Over-dehydrated or inadequately fixed tissue shows nonspecific adherence to the polymer cap. The problem is solved by reduce dehydration gradient staining time or increasing fixative treatment. In case of nonspecific adherence of the tissue, use an adhesive note (CapSure Clean-up pad or tacky adhesive note paper) in order to gently blot surplus tissue from the surface of the polymer LSM (*see* Fig. 1).

7. To prepare DPEC-treated water, add 1.5 mL of DPEC to 1 L of deionized water, and mix well by leaving the mixture on continuous stirring for 2 h. Autoclave the mixture twice before use to remove traces of DPEC. To treat tips and Eppendorf with DPEC, add DPEC-treated non-autoclaved water to the container having the mentioned plasticware, and stir overnight. The next day, remove DPEC-treated water and autoclave the plasticware twice.

8. Vortex the sample, and incubate on ice until all the samples are processed. Rinse the pestle and motor with another 150 μL of the phenol and chloroform in order to maximize the RNA yield.

9. Decreased yield of RNA may be due to insufficient dewaxing of the samples or insufficient number of cells microdissected.

References

1. Sheridan C, Depuydt P, De Ro M, Petit C, Van Gysegem E, Delaere P, Dixon M, Stasiak M, Aciksöz SB, Frossard E et al (2017) Microbial community dynamics and response to plant growth-promoting microorganisms in the rhizosphere of four common food crops cultivated in hydroponics. Microb Ecol 73:378–393
2. Lawlis T, Parker J, Green A (2017) The effects of auxin on plant growth to apical meristem and varying effects of light intensities. The effects of auxin on plant growth to apical meristem and varying effects of light intensities
3. Miwa H, Kinoshita A, Fukuda H, Sawa S (2009) Plant meristems: CLAVATA3/ESR-related signaling in the shoot apical meristem and the root apical meristem. J Plant Res 122:31–39
4. García-Gómez ML, Azpeitia E, Álvarez-Buylla ER (2017) A dynamic genetic-hormonal regulatory network model explains multiple cellular behaviors of the root apical meristem of *Arabidopsis thaliana*. PLoS Comput Biol 13: e1005488
5. De Tullio MC, Jiang K, Feldman LJ (2010) Redox regulation of root apical meristem organization: connecting root development to its environment. Plant Physiol Biochem 48:328–336
6. Malamy JE, Benfey PN (1997) Organization and cell differentiation in lateral roots of *Arabidopsis thaliana*. Development 124:33–44
7. Palovaara J, de Zeeuw T, Weijers D (2016) Tissue and organ initiation in the plant embryo: a first time for everything. Annu Rev Cell Dev Biol 32:47–75
8. Ji H, Wang S, Li K, Dr S, Koncz C, Li X (2015) PRL1 modulates root stem cell niche activity and meristem size through WOX5 and PLTs in *Arabidopsis*. Plant J 81(3):399–412
9. Fisher AP, Sozzani R (2016) Uncovering the networks involved in stem cell maintenance and asymmetric cell division in the *Arabidopsis* root. Curr Opin Plant Biol 29:38–43
10. Bizet F, Hummel I, Bogeat-Triboulot M-B (2014) Length and activity of the root apical meristem revealed in vivo by infrared imaging. J Exp Bot 66:1387–1395
11. Sozzani R, Iyer-Pascuzzi A (2014) Postembryonic control of root meristem growth and development. Curr Opin Plant Biol 17:7–12
12. Müller J, Toev T, Heisters M, Teller J, Moore KL, Hause G, Dinesh DC, Katharina B, Abel S (2015) Iron-dependent callose deposition adjusts root meristem maintenance to phosphate availability. Dev Cell 33:216–230
13. Hussain A, Hasnain S (2011) Interactions of bacterial cytokinins and IAA in the rhizosphere may alter phytostimulatory efficiency of rhizobacteria. World J Microbiol Biotechnol 27:2645–2654
14. Pérez-Flores P, Valencia-Cantero E, Altamirano-Hernández J, Pelagio-Flores R, López-Bucio J, García-Juárez P, Macías-Rodríguez L (2017) *Bacillus methylotrophicus* M4-96 isolated from maize (*Zea mays*) rhizoplane increases growth and auxin content in *Arabidopsis thaliana* via emission of volatiles. Protoplasma:1–13
15. Wang J, Zhang Y, Li Y, Wang X, Nan W, Hu Y, Zhang H, Zhao C, Wang F, Li P et al (2014) Endophytic microbes *Bacillus* sp. LZR216-regulated root development is dependent on polar auxin transport in *Arabidopsis* seedlings. Plant Cell Rep 34:1075–1087
16. Verma SK, Kingsley K, Bergen M, English C, Elmore M, Kharwar RN, White JF (2017) Bacterial endophytes from rice cut grass (*Leersia oryzoides* L.) increase growth, promote root gravitropic response, stimulate root hair formation, and protect rice seedlings from disease. Plant Soil:1–16
17. Hussain A, Ullah I, Hasnain S (2017) Microbial manipulation of auxins and cytokinins in plants. Aux Cyto Plant Biol:61–72
18. Ioio RD, Nakamura K, Moubayidin L, Perilli S, Taniguchi M, Morita MT, Aoyama T, Costantino P, Sabatini S (2008) A genetic framework for the control of cell division and differentiation in the root meristem. Science 322:1380–1384
19. Martin LBB, Nicolas P, Matas AJ, Shinozaki Y, Catalá C, Rose JKC (2016) Laser microdissection of tomato fruit cell and tissue types for transcriptome profiling. Nat Protoc 11:2376–2388

Chapter 7

Live Imaging of *Arabidopsis* Axillary Meristems

Bihai Shi, Hongli Wang, and Yuling Jiao

Abstract

Axillary meristems (AMs) are established postembryonically at the leaf axils and can develop into lateral branches. The initiation of AMs establishes new growth axis and is of primary importance for understanding plant development. Understanding plant development requires live imaging of morphogenesis and gene expression. However, AMs are embedded in the leaf axil, making it challenging to perform live imaging. In this chapter, we describe how to prepare and culture *Arabidopsis thaliana* leaves in vitro, to perform one-time or time-lapse imaging of AM initiation with a confocal microscope.

Key words Axillary meristem, In vitro culture, Confocal, Live imaging, *Arabidopsis*

1 Introduction

Plants continuously form new tissues and organs throughout their life span in a modular manner. Each module, termed as phytomer, includes a leaf, an axillary meristem (AM) at the leaf axil, and a segment of internode. AMs form lateral buds and have the potential to further develop into lateral branches. The activity of AMs thus determines the plant architecture and crop yield [1].

Development is a dynamic progress involving cell division, spatiotemporal gene expression, and protein localization. Traditional techniques, such as in situ hybridization and analysis of transgenic plants harboring nonfluorescent molecular markers, use dead, fixed samples and lack proper spatiotemporal resolution. The development of confocal live-imaging technique opens a new way to investigate developmental processes, such as AM initiation in live samples at the cellular resolution. In this chapter, we present a detailed live-imaging method to study AM initiation using an upright confocal microscope. This method has been used to capture the expression dynamics of the auxin sensor DII-Venus and meristem marker genes during AM initiation [2–4]. This chapter also covers preparation and in vitro culture of leaves for imaging, including important points and tips.

Muhammad Naseem and Thomas Dandekar (eds.), *Plant Stem Cells: Methods and Protocols*, Methods in Molecular Biology, vol. 2094, https://doi.org/10.1007/978-1-0716-0183-9_7, © Springer Science+Business Media, LLC, part of Springer Nature 2020

2 Materials

1. Confocal laser scanning microscope (CLSM): An upright confocal microscope is preferred, but an inverted microscope could also be used [5, 6].
2. Water dipping lens: 40× objective lens with a large working distance (e.g., Nikon NIR Apo 40×, 0.8 NA, 3.5 mm working distance).
3. Stereomicroscope: For tissue dissection, a magnification of at least 50× is suggested.
4. Tweezers sterilizer (e.g., Keller Steri 250 dry beads sterilizer; Keller, Swiss) (*see* **Note 1**).
5. Laminar flow hood: For time-lapse imaging, all the steps mentioned in the Methods section before confocal imaging are performed in a hood to avoid contamination.
6. Tweezers: Tweezers with fine tips are used to transfer and dissect seedlings and leaves.
7. Hypodermic sterile insulin syringes: Sterile syringe needle tips act as sharp blades to cut leaves from the seedling.
8. Murashige and Skoog (MS) medium: 1/2× MS basal salt mixture with vitamins, 2% sucrose, pH = 5.8, 0.8% agarose, autoclaved at 121 °C for 20 min.
9. MS plates for seed germination: Round glass petri dishes (~10 cm in diameter and ~2 cm in depth) filled with ~0.7 cm MS medium in depth.
10. In vitro culture medium: 1× MS basal salt mixture with vitamins, 2% sucrose, 0.0005% (w/w) folic acid, 0.01% (w/w) myo-inositol, pH = 6.0, 0.3% phytagel, autoclaved at 113 °C for 15 min (*see* **Note 2**).
11. Dissecting plates: Round glass petri dishes (~10 cm in diameter and ~2 cm in depth) filled with autoclaved 3% agarose (*see* **Note 3**).
12. Imaging plates: For time-lapse imaging, use round plastic petri dishes (~6 cm in diameter and ~1.5 cm in depth; Corning Life Sciences, Corning, NY, USA) with ~0.7 cm MS medium in depth. After cooling down, top up with a layer (~0.3 cm in depth) of autoclaved 1% agarose to minimize contamination. For one-time imaging, use dishes with 1 cm 1% agarose instead.
13. In vitro culturing plates: Round glass petri dishes with ~1 cm in depth in vitro culture medium.
14. Plants for imaging: 15-d-old *Arabidopsis thaliana* seedlings grown on MS plates under short-day conditions (8-h light at

21 °C, and 16-h dark at 19 °C) (*see* **Note 4**). We have used L*er*, Col-0 and Ws, and obtained comparable results. We expect other ecotypes to work as well.

15. FM4–64 solution: For cell membrane staining (Thermo Fisher Scientific, Waltham, MA, USA; stock concentration, 20 mg/mL; working concentration, 50 μg/mL) [7].
16. Illumination incubator (e.g., Sanyo, MLR-351H, Osaka, Japan).
17. Autoclaved ddH_2O.
18. Air-permeable tape (e.g., Scotch filter tape, 3 M; St. Paul, MN, USA).

3 Methods

1. Use tweezers to make a small vertical hole at the center of a dissecting plate (*see* **Note 5**).
2. Transfer an intact seedling from MS plate to the dissecting plate near the hole. Using stereomicroscope, transplant the seedling into the hole with the petiole base just touching the agarose surface (*see* Fig. 1a and **Note 6**).
3. Dig out a small piece of wedge-shaped agarose from the periphery of the dissecting plate, and use it to securely wedge the seedling into the hole. Make sure the seedling sits stable (*see* **Note** 7).
4. To detach the first pair of true leaves (i.e., P_{11} and P_{10}), first press the leaf petiole down while gently dragging it left and right to gradually reveal leaf axil and abaxial incision line (the junction between the petiole basal part and the hypocotyl). Use a syringe needle to cut along flanks of the abaxial incision line, and keep dragging outward gently until the leaf is detached (*see* Fig. 1b, c, and **Note 8**). If necessary, cut the vasculature linkage between the detaching leaf and the hypocotyl. Float the detached leaf on the in vitro culture plate with the petiole basal part touching the medium.
5. To dissect the younger leaves (P_9–P_7), use tweezers to gently press the leaf petiole or blade down to reveal the leaf axil. Use a syringe needle to carefully remove two stipules lying on the leaf axil, and then detach the leaf. Collect leaves from one seedling, and transfer to an in vitro culture plate for staining and imaging (*see* Fig. 1d).
6. Transfer one leaf to an imaging plate, and apply one drop of FM4–64 dye solution to the petiole basal part, where AM progenitor cells reside. Check under stereomicroscope to ensure that the basal part is immersed completely in the solution for homogeneous staining (*see* **Note 9**).

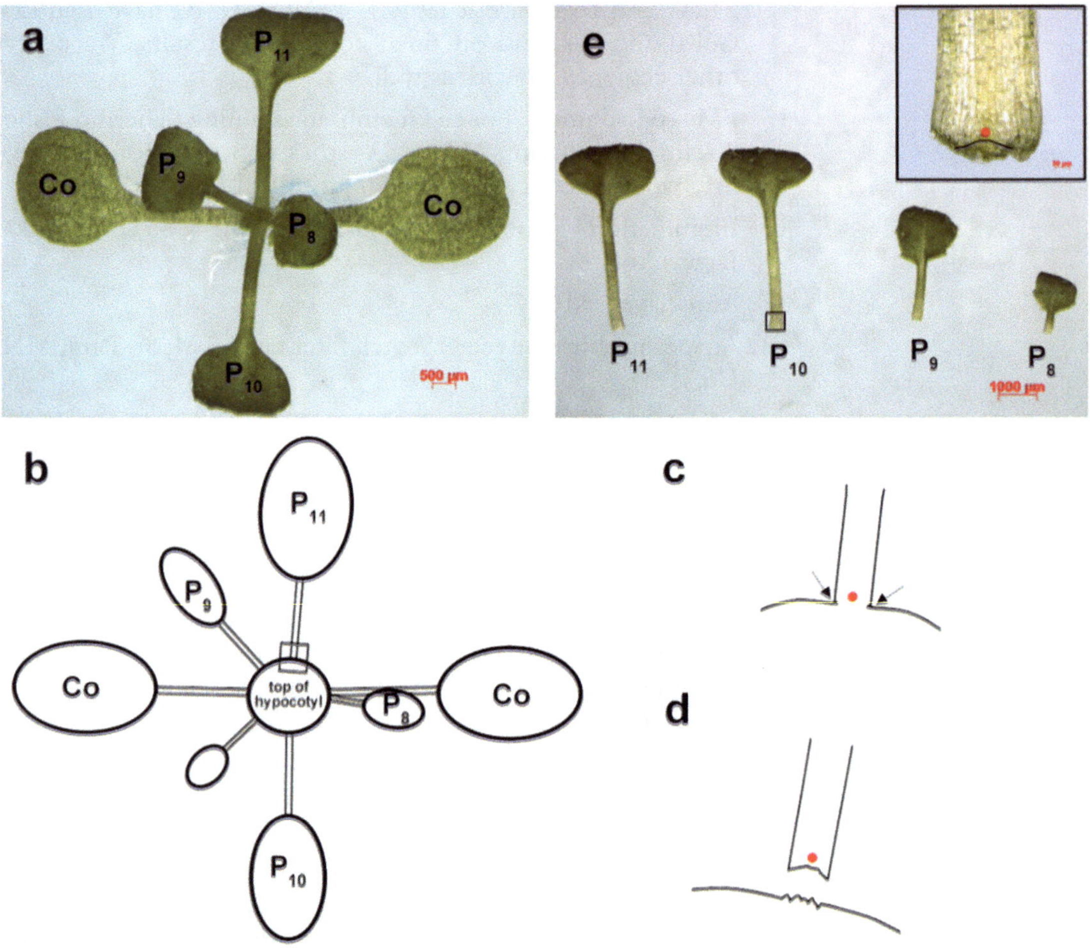

Fig. 1 Detaching leaves for AM imaging. (**a**) A 16-d-old Col-0 seedlings grown under short-day conditions with two cotyledons (Co) and leaves P_{11} to P_7 visible by naked eyes sit in a dissecting plate. (**b**) Diagram of (**a**) to highlight the leaf axils, one of which is enclosed with a black rectangle. (**c**) Schematic of detaching a leaf from the seedling. Arrows indicate flanks of the leaf boundary. The red dot marks AM progenitor cells. (**d**) Schematic of a detached leaf with AM progenitor cells (the red dot) undamaged. (**e**) Leaves (from P_8 to P_{11}) detached from the same seedling floating on an in vitro culture plate. The insert shows corresponding magnified basal part (including the leaf axil) of the P_{10} indicated by a black box. The red dot marks AM progenitor cells, and the black line below indicates the boundary. Bars = 500 μm in (**a**), 1000 μm in (**b**), and 50 μm in the insert of (**b**)

7. After 20 min, transfer the stained leaf to another imaging plate. Under a stereomicroscope, insert the leaf blade into the center of the medium, and adjust the angle of inserted leaf blade until the leaf axil harboring meristem cells is horizontal and facing upward (*see* Fig. 2a, b and **Note 10**).
8. Cover the exposed petiole basal part in the imaging plate with sterile ddH_2O. Remove any air bubbles around the petiole tip by pipetting, and then cover the plate with a lid (*see* **Note 11**).

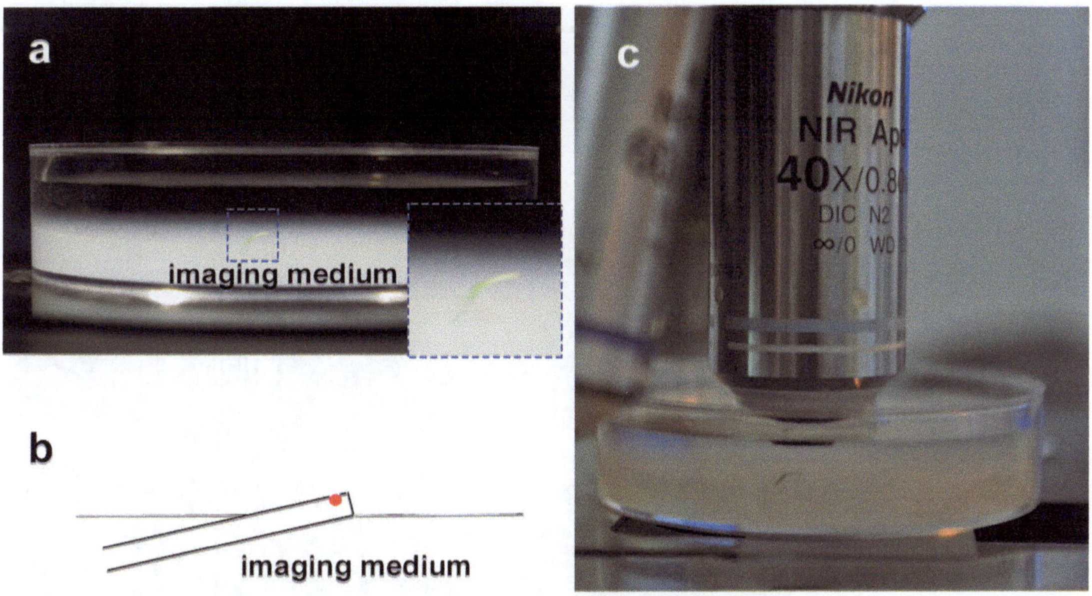

Fig. 2 Preparation of detached leaves for AM imaging. (**a**) A leaf blade inserting in an imaging medium with the leaf axil facing upward, positioned horizontally and covered with water. (**b**) Diagram of (**a**) to highlight the insertion angle. (**c**) View of an imaging plate with a leaf insert in the central part as in (**a**) on a CLSM stage. A 40× water dipping lens is used

9. Position the inserted leaf blade under water dipping lens of CLSM, and roughly aim it to the leaf axil by adjusting the stage. Then, raise the microscope stage until the objective lens immerses into water, but not yet touching the leaf axil (*see* Fig. 2c). Carefully look from all sides to ensure there are no bubbles attaching to the objective lens. Otherwise, lower the stage and redo the immersion.
10. Under epifluorescence illumination, position the leaf axil exactly within the field of the objective lens by further adjusting the stage. Follow the acquisition procedures suggested by the CSLM manufacture to obtain 3D images of the leaf axil (*see* Fig. 3).
11. After imaging, lower the stage, take out the plate, and close its lid to take it back to the hood. After pouring out the water with the help of a pipette, gently drag the leaf out of the hole and transfer it back to a culturing plate. Make sure the petiole basal part touches the medium surface so that the leaf is kept moist and unstressed.
12. Seal the culturing plate with air-permeable tape, and put it in an illumination incubator with short-day conditions.
13. For the following time points (usually at an interval of 24 h), unseal the culturing plate in a sterilized hood and repeat **steps 6–12**. For untreated wild-type plants, it takes 4–5 days to see AM initiation with its own primordia forms.

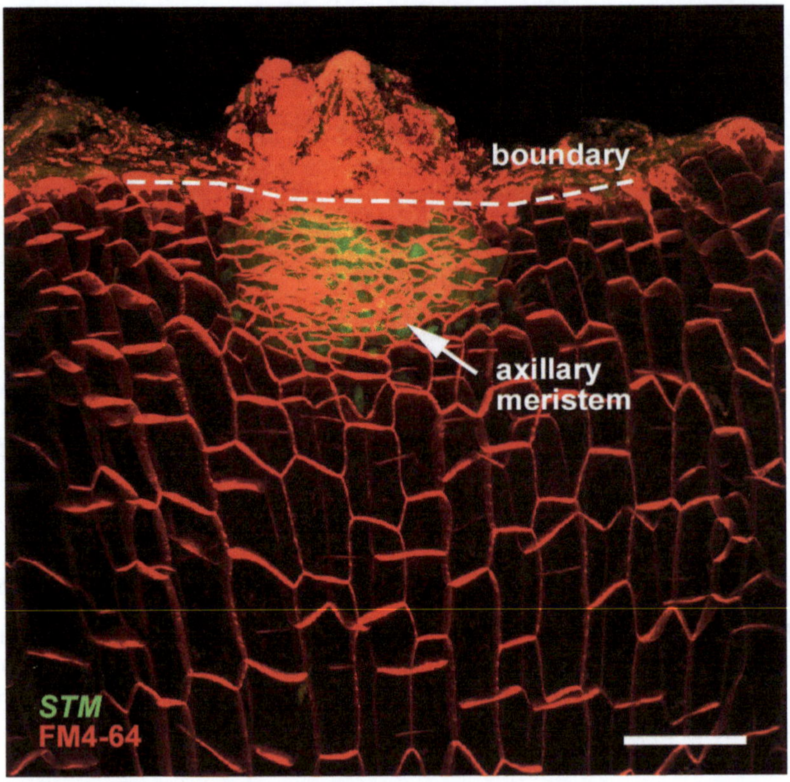

Fig. 3 Maximum intensity projection view of confocal z-stacks of a FM4–64 stained (red) *pSTM::STM-Venus* (green) leaf basal part from a live P_8 2d after in vitro culture. The arrow indicates the AM, and the white dotted line above highlights the boundary. Bar = 50 μm

4 Notes

1. A sterilizer is convenient, but not necessary. Flame sterilization with an alcohol burner also works.
2. Do not over autoclave phytagel.
3. Round edges of glass dishes make it more comfortable to dissect plants inside.
4. Short-day condition is required to avoid precocious AM initiation and axillary bud formation.
5. Stabbing holes in the center of plates provide a convenient angle for dissection.
6. To avoid damaging leaf blade, use tweezers to hold the junction between leaf petioles and the stem to transfer seedlings. Under a stereomicroscope, put the entire roots and hypocotyl into the hole and the stem base above the agarose.

7. It is crucial to ensure the seedling firmly secured in the agarose for dissection. A proper sized piece of wedge-shaped agarose may help.
8. AM progenitor cells are small in size and lie right next to the boundary, i.e., the junction between the petiole base and the stem. Therefore, try to keep the boundary intact to avoid damaging meristematic cells.
9. Once the imaging medium surface is wet, change to a new one and air dry the former one in a hood. The FM4-64 dye will be absorbed into a wet imaging medium surface.
10. It is helpful to make a sloppy path in the medium using tweezers to insert a leaf blade. It is important to insert a leaf blade in the center of a plate to provide enough space for the objective lens.
11. This is to minimize contamination when a plate is taken out of the hood for imaging.

Acknowledgments

We thank Max Bush for schematics and Muhammad Sajjad for proofreading. Research in the authors' laboratory is supported by the National Natural Science Foundation of China grants 31430010, 31861130355, and 31825002, the Key Project of the Frontier Science of the Chinese Academy of Sciences grant ZDBS-LY-SM012, and a Royal Society Newton Advanced Fellowship (NAF/R1/180125).

References

1. Schmitz G, Theres K (2005) Shoot and inflorescence branching. Curr Opin Plant Biol 8:506–511
2. Wang J, Tian C, Zhang C, Shi B, Cao X, Zhang T-Q, Zhao Z, Wang J-W, Jiao Y (2017) Cytokinin signaling activates *WUSCHEL* expression during axillary meristem initiation. Plant Cell 29:1373–1387
3. Shi B, Zhang C, Tian C, Wang J, Wang Q, Xu T, Xu Y, Ohno C, Sablowski R, Heisler MG, Theres K, Wang Y, Jiao Y (2016) Two-step regulation of a meristematic cell population acting in shoot branching in *Arabidopsis*. PLoS Genet 12:e1006168
4. Wang Y, Wang J, Shi B, Yu T, Qi J, Meyerowitz EM, Jiao Y (2014) The stem cell niche in leaf axils is established by auxin and cytokinin in *Arabidopsis*. Plant Cell 26:2055–2067
5. Nimchuk ZL, Perdue TD (2017) Live imaging of shoot meristems on an inverted confocal microscope using an objective lens inverter attachment. Front Plant Sci 8:773
6. Prunet N, Jack TP, Meyerowitz EM (2016) Live confocal imaging of *Arabidopsis* flower buds. Dev Biol 419:114–120
7. Reddy GV, Heisler MG, Ehrhardt DW, Meyerowitz EM (2004) Real-time lineage analysis reveals oriented cell divisions associated with morphogenesis at the shoot apex of *Arabidopsis thaliana*. Development 131:4225–4237

Chapter 8

Molecular Modeling of the Interaction Between Stem Cell Peptide and Immune Receptor in Plants

Muhammad Naseem, Mugdha Srivastava, Ozge Osmanoglu, Jibran Iqbal, Fares M. Howari, Fatima A. AlRemeithi, and Thomas Dandekar

Abstract

Molecular docking enables comprehensive exploration of interactions between chemical moieties and proteins. Modeling and docking approaches are useful to determine the three-dimensional (3D) structure of experimentally uncrystallized proteins and subsequently their interactions with various inhibitors and activators or peptides. Here, we describe a protocol for carrying out molecular modeling and docking of stem cell peptide CLV3p on plant innate immune receptor FLS2.

Key words FLS2, CLV3p, Docking, Modeling, Structure prediction, Protein peptide interaction

1 Introduction

Innate immunity in plants is triggered through the pattern recognition receptors (PRRs) in response to microbe-associated molecular patterns (MAMPs) [1, 2] which provides the first line of inducible defense against pathogen attack. Flagellin-sensing 2 (FLS2) is a leucine-rich immune receptor in plants which responses to bacterial flagellin by enabling the defense and signaling pathways for pathogen resistance [3–5]. FLS2 is sensitive to the conserved epitopic region (flg22) in the flagellar protein of various pathogenic as well as nonpathogenic bacteria [6, 7]. FLS2 is expressed in different parts of plant body including shoot apical meristem (SAM), and flagellin-FLS2 signaling provides immune protection in plants after infection. It has been shown that a 12-amino-acid CLV3p triggered similar responses as flg22 (the conserved 22-amino-acid peptide of bacterial flagellin) in mesophyll protoplasts. flg22 and CLV3p, but not CLV3p lacking the last histidine residue, activate similar mitogen-activated protein

Authors Muhammad Naseem and Mugdha Srivastava contributed equally to this work.

Muhammad Naseem and Thomas Dandekar (eds.), *Plant Stem Cells: Methods and Protocols*, Methods in Molecular Biology, vol. 2094, https://doi.org/10.1007/978-1-0716-0183-9_8, © Springer Science+Business Media, LLC, part of Springer Nature 2020

kinase (MAPK) activities in the experiments performed by Lee et al. in 2011 [4]. Also, it was found that different modified variants of CLAVATA3 (CLV3) peptides have strikingly different activities in immune response gene activation mediated by the FLS2 and CLAVATA2 (CLV2) receptors, respectively. Interestingly, findings of other groups actively working on innate immunity in plants questioned the CLV3p-mediated immunity through its binding to FLS2 and the subsequent signaling events such as ROS production and mitogen-activated protein kinase (MAPK) activities as well as the expression of defense marker genes [8, 9]. To assess the binding dynamics of three variants of CLV3 peptides to immune receptor FLS2, we used the structural modeling and molecular docking approaches.

Several methods such as X-ray crystallography, NMR spectroscopy, cryo-electron microscopy, and others are available for the structural characterization of proteins; however, due to the technical difficulties and labor intensiveness of these methods, the number of proteins is frequently modeled by computational methods to annotate the biological function of a protein molecule whose structure is not available in Protein Data Bank (PDB) [10]. Computational methods under three broad categories, comparative modeling (CM), threading, and ab initio modeling [11], are frequently used to determine the 3D structure of proteins or peptides. The first two methods are template based, and in the absence of complete query coverage, the full-length structure cannot be determined. The ab initio modeling algorithm is free of templates, and the structure is built from scratches. Here, we define the protocol to use multi-template modeling, a variant of comparative modeling for structure prediction of CLV3p and derivatives, and further docking with ZDOCK to study the molecular interactions between FLS2 and CLV3p. User manuals of the used software can be used as reference in addition to this chapter (*see* **Notes 1** and **2**).

This chapter is intended for general biology audience that is at beginner level with respect to modeling and docking. Basic skills in using Windows and Linux operating systems are expected. We show the real example of these approaches by investigating the binding behavior of modeled CLV3p and its variants onto immune receptor FLS2.

2 Materials

2.1 Required Computing Hardware

1. Personal computer running on a Linux (preferably) operating system.

2.2 Required Software

1. Modeller 9.11 or higher version from https://salilab.org/modeller/download_installation.html. Modeller [12] is a freeware

and accessible for Windows, Mac, and Linux operating systems. Key can be obtained from the Modeller website under the academic user license.

2. We recommend at least Python 2.3 or higher version (https://www.python.org/downloads/) compatible to the downloaded version of Modeller.
3. Discovery Studio® platform for mutation modeling (http://accelrys.com/products/collaborative-science/biovia-discovery-studio/).
4. Web servers: BLAST (https://blast.ncbi.nlm.nih.gov/Blast.cgi?PAGE=Proteins) for sequence similarity searches [13], KoBaMIN (http://chopra-modules.science.purdue.edu/modules/kobamin/html/) for energy minimizations [14], PROCHECK (http://servicesn.mbi.ucla.edu/PROCHECK/) for structure validation [15], and PDBsum (http://www.ebi.ac.uk/thornton-srv/databases/cgi-bin/pdbsum/GetPage.pl?pdbcode=index.html) for analyzing the chemical bonds in docked complexes [16].
5. The docking software ZDOCK can be obtained from http://zdock.umassmed.edu/software/download/ [17, 18] under academic user license. This distribution includes an executable file (ZDOCK) of the ZDOCK program, PDB processing file (mark_sur, uniCHARMM, block.pl), and auxiliary files (create.pl, create_lig) to create predicted complex structures from a ZDOCK output. ZDOCK is available for downloads for Linux, Mac, and Unix platforms. We recommend version at least 3.0 for the docking and energy calculations.
6. ZRANK program is available at http://zdock.umassmed.edu/software/download/ [19] for reranking sets of initial-stage ZDOCK docking predictions using an optimized energy-based function.

2.3 Required Data

1. Crystal structure of FLS2 (PDB entry 4MN8) from Protein Data Bank (PDB) database [20].
2. Peptide sequence of CLV3p from literature [21].

3 Methods

3.1 Molecular Modeling of CLV3 Peptide

1. Perform similarity search of the amino acid sequence using by BLASTp at NCBI. Put the CLV3p sequence in the textbox under "Enter Query Sequence," and select database "Protein Data Bank proteins (pdb)" in drop-down menu under "Choose Search Set." Click on BLAST, and the result window will appear in couple of minutes. The server will automatically tune the algorithm parameters for the BLAST search using

short peptide sequence input such as adjustment of word size and gap penalty parameters (*see* **Notes 1–3**).

2. By manual inspection of the resulting alignment, the templates can be selected with lower e-value and good alignment to satisfy 100% query coverage.
3. Use Modeller for multi-template modeling. This uses the restraints from the geometry of selected template to model the query peptide (CLV3p) structure. Please refer to the notes section for short example of multi-template modeling. Readers can also refer to https://salilab.org/modeller/tutorial/advanced.html for further detailed explanation.
4. Structure validation and iterative refinement are important steps that should be performed before using the structure. The predicted model geometry can be refine using the KoBaMIN web server, and the quality assessment of the refined energy-minimized model can be performed by the inspection of the Psi/Phi torsion angles of residues in the quadrants of the Ramachandran plot obtained from PROCHECK analysis (*see* **Notes 2** and **3**).

3.2 Mutation Modeling

1. To analyze the effect of the amino acid residue histidine on the binding affinity with FLS2, build different mutants using the "build-mutant" protocol of Discovery Studio. The protocol optimizes all the atoms of mutated residues using a scoring function which includes molecular mechanics energy terms for dihedral angles as well as bond angles, peptide bond planarity, bond distances, Lennard-Jones potential for nonbonded interaction, homology-derived restraints for main-chain and side-chain dihedral angles and the statistical potential for non-bonded atom interactions extracted from a large set of known protein structures [22].
2. Build the CLV3p variants by removing the last histidine in each and the subsequent energy minimization. Substituting arginine1, proline9, and histidine11 with alanine, respectively, to build the other mutants. Since the mutation of a single amino acid residue may significantly affect the conformation of the residue around 4.5 angstrom [22, 23], it is essential to optimize the confirmation of neighboring residues that lie within a specified cutoff radius of 4.5A° after mutating the specific residues.
3. Subject all the mutant models to energy minimization to improve the stereochemistry of models with a knowledge-based force field using the KoBaMIN server.
4. Estimate the quality of refined mutant models with PROCHECK.

5. Repeat **steps 3** and **4** until getting the best structure of all the mutant models.

3.3 Docking Analysis

1. Preprocess the PDB format files of receptor and ligand (peptide) with mark_sur program which measures the amount of accessible surface area (ASA) of each atom using a water probe of radius 1.4 A°. If an atom has as ASA of more than 1 $A^{o\ 2}$, it will be marked as a surface atom. Force field uniCHARMM, supplied with ZDOCK program, should be in the same folder from where mark_sur will be run. It attributes to pre-docking minimization, add missing atoms and polar hydrogens to the proteins (*see* **Note 3**).

```
>> mark_sur clv_ligand.pdb clv_ligand_output.pdb
```

```
>> mark_sur FLS2_receptor_blocked.pdb FLS2_receptor_output.
pdb
```

2. Analyze the interaction of flg22 and FLS2 using PDBsum, and block the residues not participating in the interaction during the docking analysis. These blocked residues are treated as a special type as they are not likely to be located in the binding site of CLV peptides. Therefore, only the penalty part of the pairwise shape complementarity (PSC) score is applied to them during the calculation. As a result, the complex conformations with the specified residues in the interface receive lower scores than other conformations. This blocking procedure can improve docking performance significantly. The program needs a user-defined list file with the amino acid number (one per line) of the receptor not likely to be the part of the binding site (Fig. 1).

```
>> block.pl FLS2_receptor_output.pdb residue_to_block.txt
```

```
Save the screen output file as FLS2_receptor_output_blocked.
pdb.
```

3. Perform docking analysis between the different variants of CLV3p and FLS2 (Table 1). ZDOCK algorithm uses a fast Fourier transform (FFT)-based docking approach to find the 3D structure of protein complex. For each ligand rotation, it is discretized receptor to obtain the top-scoring ligand position. Docking using ZDOCK is normally performed in two stages. In the first stage, proteins are treated as rigid bodies, and 54,000 predictions are generated. In the next refinement stage, a combination of detailed scoring, energy minimization, side-chain searches, and clustering is performed on these predictions. The algorithm optimizes three parameters: shape complementarity, electrostatics, and desolvation free energy. ZDOCK uses a simple shape complementarity method called

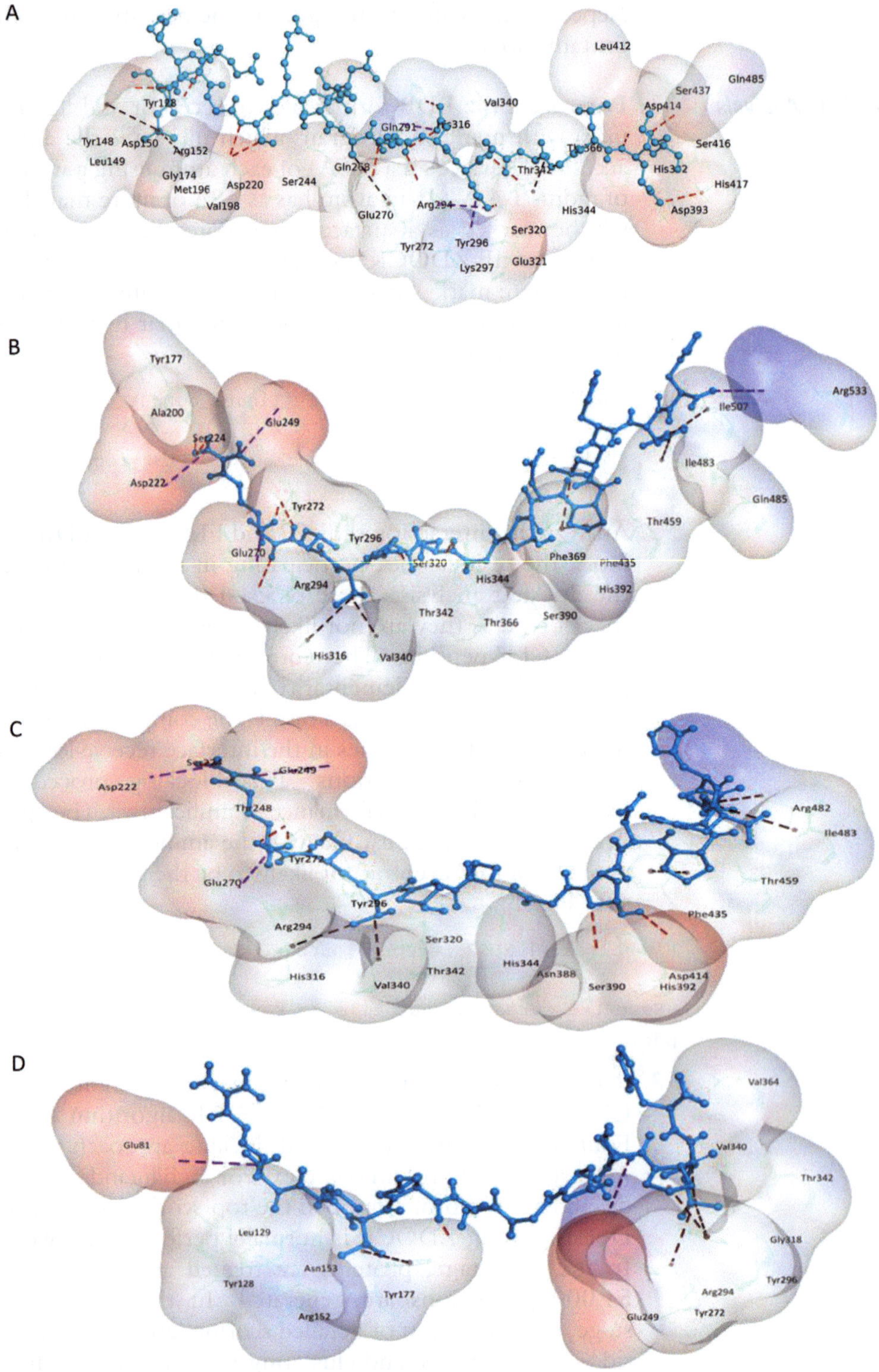

Fig. 1 Interactions between plant signaling peptide CLAVATA3p and immune receptor FLS2. (**a**) The figure depicts the hydrogen and electrostatic and hydrophobic interactions for the FLS2-flg22 complex (PDB entry 4MN8). The ball and stick blue-colored structure represents the peptide flg22, and the surface structure

Table 1
Docking score for all the different CLAVATA3p variants

S. no.	Nature of interaction	ZDOCK score
1.	13aa-CLV3p-FLS2	−142.576
2.	13aa-MCLV3p-FLS2	−156.276
3.	12aa-CLV3p-FLS2	−163.871
4.	12aa-MCLV3p-FLS2	−173.431
5.	12aa-MCLV3p(Argl → Ala)-FLS2	−141.607
6.	12aa-MCLV3p(Pro9 → Ala)-FLS2	−150.872
7.	12aa-MCLV3p(His11 → Ala)-FLS2	−144.402
8.	12aa-MCLV3p-FLS2*	−149.73
9.	11aa-MCLV3p-FLS2	−138.119

MCLV3p represents peptide modified at the 4 and 7 positions by mutating proline to hydroxyproline. The symbol (*) represents blind docking, while all other interactions represent site-specific docking

PSC [24]. PSC has been shown to yield better results than the common grid-based shape complementarity (GSC) method and has been proven effective both against a docking benchmark 7 and in several rounds of the CAPRI experiment [25]. It rewards contiguous surface patches at the binding site and implicitly accounts for the curvature of the binding surface. Both the steps are repeated to cover all ligand rotations in three dimensions, if necessary. The command involves 54,000 (-N) iterations for 6o (-D) sampling. The -N option asks for 54,000 docking decoys.

```
>> ZDOCK -D 6 -N 54000 -R FLS2_receptor_output_blocked.pdb -L
clv_ligand_output.pdb -o ZDOCK_receptor_ligand_complex_54000.
out
```

where ZDOCK_receptor_ligand_complex_54000.out is the output (-o) of the ZDOCK program.

Fig. 1 (continued) represents immune receptor FLS2. The red dotted lines represent the hydrogen bond interactions; purple represents the electrostatic, while brown represents the hydrophobic interactions. (**b**) FLS2-12-aa CLV3p interaction. The ball and stick blue-colored structure represents the peptide 13-aa CLV3p (CLAVATA3p with 13 amino acid residues), and the surface structure represents FLS2 receptor. Only interacting residues are shown here. (**c**) FLS2-12-aa CLV3p interaction. The ball and stick blue-colored structure represents the peptide 12-aa CLV3p (CLAVATA3p with 12 amino acid residues), and the surface structure represents FLS2 receptor. Only interacting residues are shown here. (**d**) FLS2-11-aa CLV3p interaction. The ball and stick blue-colored structure represents the peptide (CLAVATA3p with 11 amino acid residues), and the surface structure represents FLS2 receptor. Only interacting residues are shown here

4. The ZDOCK program does not create atomic coordinates for the predicted complex structures automatically. Due to the input rotation of the receptor and the possible switching of the ligand and receptor, the ZDOCK output file format required slight modification of its header format and consequent updating of the code to create ZDOCK predictions (create_lig) from the output file. The program will generate the atomic coordinates (PDB files) of a specified complex/pose from the ligand rotational and translational matrices stored in the ZDOCK_receptor_ligand_complex_54000.out file.
   - Create a directory on your computer, and download the ZDOCK output (here ZDOCK_receptor_ligand_complex_54000.out), receptor, and ligand file to this directory which was used in ZDOCK command.
   - Copy the create_lig program to the directory with the ZDOCK output file.
   - Run the script "create.pl" from this directory. It will generate all the predicted complexes in the ZDOCK output file and name them "complex.X" where "X" is the prediction number.

```
>> create.pl ZDOCK_receptor_ligand_complex_54000.out 54000

>> ls complex* > listfile

>> sort -V listfile > listfile2
```

5. As all the peptide was docked with receptor FLS2 and the predicted 54,000 poses/complexes, it is further crucial to rank the docking poses. ZRANK (Zlab Rerank) quickly and accurately reranks the rigid body docking results from ZDOCK by utilizing a scoring function that can be rapidly computed and effectively employed to discriminate hits from non-hits. The scoring function is a linear weighted sum of van der Waals attractive and repulsive energies, electrostatics short- and long-range attractive and repulsive energies, and desolvation energy.

```
>> zrank -R listfile2

>> sort -n -k 2 listfile2.zr.out > result_clv11ds_not_modi-
fied_54000
```

6. Analyze the best predicted complexes by ZRANK for the presence of bonds using PDBsum server.
7. Repeat the **steps 1**, **3**, **4**, **5**, and **6** for all the created clv peptide mutants.

4 Notes

1. Multi-template modeling can be performed by creating an alignment file in the PIR format and saved with an extension of ali. The templates used for the modeling of CLV3p are 1WQUA|A, 2D9V|A, 2LAK|A, 4EL8|A, 4GNK|B, 4J11|A, and 4MX2|A. The PIR format for the alignment file of clavatal3p is like the following: >P1;clavatal3psequence:clavatal3p::::::0.00: 0.00 RTVPSGPDPLHHH* >P1;1WQU structureN:1WQU.pdb:105:A:111:A:FES SH2 domain: human:-1.00:-1.00 RAVPSGP* >P1;2D9V structureN:2D9V.pdb:122:A:127:A: PH domain of Pleckstrin homology domain-containing protein family B member: mouse:-1.00:-1.00 TVPSGP*.
2. The first line of each sequence entry specifies the protein code after the >P1; line identifier. The second line of each entry contains information necessary to extract atomic coordinates of the segment from the original PDB coordinate set. The fields in this line are separated by colon characters, ":". The fields are as follows: Field 1 describes if the structure is available and is solved by which method, structureX represents X-ray; structureN, NMR; structureM, model; sequence as sequence. Only structure is also a valid value. Field 2 is the pdb code. Fields 3–6 are the residue identifiers, for the first (fields 3–4) and last residue (fields 5–6) of the sequence in the subsequent lines along with their chain ids. Field 7 is protein name which is optional. Field 8 is source of the protein and is an optional entry. Fields 9 and 10 are resolution and R-factor of the crystallographic analysis. Both these entries are optional. Each entry ends with an asterisk (*).
3. The next step is to build the model. The model of interest can be created by changing the respective arguments in the following standard file model: multiple.pyComparative modeling with multiple templates from Modeller import * # Load standard Modeller classes from modeller.automodel import * # Load the automodel class log.verbose() # request verbose output env = environ() # create a new MODELLER environment to build this model in # directories for input atom files env.io.atom_files_directory = ['.', '../atom_files'] a = automodel (env, alnfile = 'align-multiple.ali', # alignment filename knowns = ('1WQUA|A', '2D9V|A', '2LAK|A', '4EL8|A', '4GNK|B', ' 4J11|A', '4MX2|A'), # codes of the templates sequence = 'CLV3p') # code of the target a.starting_model= 1 # index of the first model a.ending_model = 1 # index of the

last model # (determines how many models to calculate) a. make() # do the actual comparative modeling.

Acknowledgments

We thank the German Research Foundation (DFG) for funding (TR124/B1) to TD and start-up grant (R18045) by Zayed University to MN and UAE Space Agency grant (EU1804) to FMH.

References

1. Jones JDG, Dangl JL (2006) The plant immune system. Nature 444:323–329. https://doi.org/10.1038/nature05286
2. Newman M-A, Sundelin T, Nielsen JT, Erbs G (2013) MAMP (microbe-associated molecular pattern) triggered immunity in plants. Front Plant Sci 4:139. https://doi.org/10.3389/fpls.2013.00139
3. Boller T, Felix G (2009) A renaissance of elicitors: perception of microbe-associated molecular patterns and danger signals by pattern-recognition receptors. Annu Rev Plant Biol 60:379–406. https://doi.org/10.1146/annurev.arplant.57.032905.105346
4. Lee H, Chah O-K, Sheen J (2011) Stem-cell-triggered immunity through CLV3p-FLS2 signalling. Nature 473:376–379. https://doi.org/10.1038/nature09958
5. Sun Y, Li L, Macho AP et al (2013) Structural basis for flg22-induced activation of the Arabidopsis FLS2-BAK1 immune complex. Science 342:624–628. https://doi.org/10.1126/science.1243825
6. Felix G, Duran JD, Volko S, Boller T (1999) Plants have a sensitive perception system for the most conserved domain of bacterial flagellin. Plant J 18:265–276
7. Meziane H, VAN DER Sluis I, VAN Loon LC et al (2005) Determinants of Pseudomonas putida WCS358 involved in inducing systemic resistance in plants. Mol Plant Pathol 6:177–185. https://doi.org/10.1111/j.1364-3703.2005.00276.x
8. Mueller K, Chinchilla D, Albert M et al (2012) Contamination risks in work with synthetic peptides: flg22 as an example of a pirate in commercial peptide preparations. Plant Cell 24:3193–3197. https://doi.org/10.1105/tpc.111.093815
9. Segonzac C, Nimchuk ZL, Beck M et al (2012) The shoot apical meristem regulatory peptide CLV3 does not activate innate immunity. Plant Cell 24:3186–3192. https://doi.org/10.1105/tpc.111.091264
10. Srivastava M, Gupta SK, Abhilash PC, Singh N (2012) Structure prediction and binding sites analysis of curcin protein of Jatropha curcas using computational approaches. J Mol Model 18:2971–2979. https://doi.org/10.1007/s00894-011-1320-0
11. Zhang Y (2009) Protein structure prediction: is it useful? Curr Opin Struct Biol 19:145–155. https://doi.org/10.1016/j.sbi.2009.02.005
12. Sali A, Blundell TL (1993) Comparative protein modelling by satisfaction of spatial restraints. J Mol Biol 234:779–815. https://doi.org/10.1006/jmbi.1993.1626
13. Altschul SF, Madden TL, Schäffer AA et al (1997) Gapped BLAST and PSI-BLAST: a new generation of protein database search programs. Nucleic Acids Res 25:3389–3402
14. Rodrigues JPGLM, Levitt M, Chopra G (2012) KoBaMIN: a knowledge-based minimization web server for protein structure refinement. Nucleic Acids Res 40: W323–W328. https://doi.org/10.1093/nar/gks376
15. Laskowski RA, Rullmannn JA, MacArthur MW et al (1996) AQUA and PROCHECK-NMR: programs for checking the quality of protein structures solved by NMR. J Biomol NMR 8:477–486
16. Laskowski RA, Hutchinson EG, Michie AD et al (1997) PDBsum: a web-based database of summaries and analyses of all PDB structures. Trends Biochem Sci 22:488–490
17. Pierce B, Tong W, Weng Z (2005) M-ZDOCK: a grid-based approach for Cn symmetric multimer docking. Bioinformatics 21:1472–1478. https://doi.org/10.1093/bioinformatics/bti229
18. Pierce BG, Hourai Y, Weng Z (2011) Accelerating protein docking in ZDOCK using an advanced 3D convolution library. PLoS One

6:e24657. https://doi.org/10.1371/journal.pone.0024657

19. Pierce B, Weng Z (2007) ZRANK: reranking protein docking predictions with an optimized energy function. Proteins 67:1078–1086. https://doi.org/10.1002/prot.21373
20. Berman HM, Battistuz T, Bhat TN et al (2002) The protein data Bank. Acta Crystallogr D Biol Crystallogr 58:899–907
21. Betsuyaku S, Sawa S, Yamada M (2011) The function of the CLE peptides in plant development and plant-microbe interactions. Arabidopsis Book 9:e0149. https://doi.org/10.1199/tab.0149
22. Feyfant E, Sali A, Fiser A (2007) Modeling mutations in protein structures. Protein Sci 16:2030–2041. https://doi.org/10.1110/ps.072855507
23. Gupta SK, Gupta SK, Smita S et al (2011) Computational analysis and modeling the effectiveness of "Zanamivir" targeting neuraminidase protein in pandemic H1N1 strains. Infect Genet Evol 11:1072–1082. https://doi.org/10.1016/j.meegid.2011.03.018
24. Chen R, Weng Z (2003) A novel shape complementarity scoring function for protein-protein docking. Proteins 51:397–408. https://doi.org/10.1002/prot.10334
25. Janin J (2005) Assessing predictions of protein–protein interaction: the CAPRI experiment. Protein Sci 14:278–283. https://doi.org/10.1110/ps.041081905

Chapter 9

Methods to Visualize Auxin and Cytokinin Signaling Activity in the Shoot Apical Meristem

Géraldine Brunoud, Carlos S. Galvan-Ampudia, and Teva Vernoux

Abstract

Visualizing the distribution of hormone signaling activity such as auxin and cytokinins is of key importance for understanding regulation of plant development and physiology. Live imaging and genetically encoded hormone biosensors and reporters allow monitoring the spatial and temporal distribution of these phytohormones. Here, we describe how to cultivate live shoot apical meristems after dissection for observation under the confocal microscope for up to 4 days. The shoot apical meristems are maintained on an appropriate medium allowing them to grow and initiate new organs at a frequency similar to plants grown on soil. Meristems expressing hormone biosensors and reporters allows following hormone signaling activity distribution at high spatiotemporal resolution without chemical fixation, an approach that that can also be applied to follow the dynamics of expression in vivo of any fluorescent marker.

Key words Shoot apical meristem, Phytohormones biosensors and reporters, Live imaging, Confocal microscopy

1 Introduction

The plant hormones auxin and cytokinins are key regulators of growth and development. Both are small molecules, tryptophan-derived and adenine-derived, respectively, which are very difficult to detect in vivo at high spatiotemporal resolution. The shoot apical meristem (SAM) is a specialized tissue located at the tip of the growing stem in which these hormones plays a key role. The SAM contains a stem cell niche and produces most of the aboveground parts of plants, including stem, leaves, shoots, and flowers. In the SAM (as in other parts of the plant), auxin is a central regulator of patterning and organogenesis. It is actively and polarly transported, and this transport creates auxin differential distributions or gradients throughout the tissue [1]. This non homogeneous distribution in the tissue is notably essential to organ initiation that is triggered by local accumulation of auxin. Cytokinins have been shown to promote stem cell identity and maintenance in the central

Muhammad Naseem and Thomas Dandekar (eds.), *Plant Stem Cells: Methods and Protocols*, Methods in Molecular Biology, vol. 2094, https://doi.org/10.1007/978-1-0716-0183-9_9, © Springer Science+Business Media, LLC, part of Springer Nature 2020

zone of the meristem [2]. They are also regulating the timing of organ initiation at the shoot apical meristem and as a result phyllotaxis robustness [3].

Several sensors and reporters for auxin and cytokinin are currently available. The auxin signaling input can be visualized using the DII-VENUS degradation-based biosensor, or its R2D2 derivative [4, 5], that has been engineered using the degron of an Aux/-IAA protein fused to the fast maturating VENUS yellow fluorescent protein and a nuclear localization signal (NLS). Aux/-IAA proteins are degraded upon perception of auxin, and DII-VENUS fluorescence thus provides information on the combined action of auxin concentration and of auxin perception capacity in plant cells. Auxin response factors (ARFs) and *Arabidopsis* response regulators (ARRs) are transcriptional factors that activate gene expression in response to auxin and cytokinin, respectively. Synthetic reporters, consisting of multiple ARF or ARR binding sites fused to a minimal promoter and driving a reporter protein, allows following the transcriptional activity induced downstream of auxin and cytokinin (signaling output). Auxin-induced transcription can be followed using the DR5 synthetic promoter [6, 7] or the DR5v2 promoter [5], while the TCS and TCSn synthetic promoter allows following cytokinin-induced transcription [8, 9]. These different genetically encoded markers have been largely used to follow auxin distribution and cytokinin signaling at the shoot apical patterning [3, 4, 10–12].

Live imaging microscopy of genetically encoded fluorescent reporters allow following the spatiotemporal dynamics of hormone distribution at the shoot apical meristem during developmental and physiological responses [11–13]. Here we provide a detailed description of how to cultivate live shoot apical meristems after dissection for observation under the confocal microscope for up to 4 days. We show how using the spatiotemporal distribution of auxin and cytokinin activity can be followed with hormone biosensors and reporters and correlated with key developmental processes such as stem cell maintenance, organogenesis, and patterning.

As introduced above, we use *Arabidopsis* plants expressing DR5, DR5v2, and TCS promoters to follow hormone-induced transcription. We use here *DR5::3xVENUS-N7* and *DR5v2::ntdTomato* [5, 11]. The two reporters differ not only in the fluorescent protein (FP) driven by the promoter but also in the repetitive DNA sequences bound by ARF (auxin responsive elements—AuxRE) in the promoter [5]. *DR5v2* has been demonstrated to have a higher sensitivity than *DR5* in certain tissues. For cytokinin, we use the *TCS::GFP* reporter [8].

To detect rapid changes in auxin concentration and signaling activity, we use the degradation-based biosensors *DII-VENUS* and *R2D2* [4, 5]. *DII-VENUS* is driven by the 35S ubiquitous promoter, and *R2D2* consists in a single construct with *DII-n3x VENUS* and *mDII-ntdTomato* driven by two independent *RPS5a* promoters that provide a strong and ubiquitous expression in the SAM. While the DII sequence allows for a rapid degradation by auxin, this degradation is strongly inhibited by the mutation present in the *mDII* sequence. *R2D2* thus provides a ratiometric reading of the auxin-dependent degradation of DII-VENUS.

All the *Arabidopsis* lines used here are in the Col-0 ecotype except R2D2 that is in the Columbia-Utrecht (Col-utr) ecotype [14].

2 Materials

2.1 Half-MS Medium

- *Prepare medium with*:

  Half-MS medium without vitamins.

  Sucrose 1%.
- Adjust pH to 5.8 with 1 M KOH.
- Add agarose at 0.8% final.
- Autoclave the medium.
- Use to pour plates or store at RT once solidified until plates are needed.

2.2 Vitamins Stock (1000×)

Myo-inositol	5 g
Nicotinic acid	0.05 g
Pyridoxine hydrochloride	0.05 g
Thiamine hydrochloride	0.5 g
Glycine	0.1 g
Add water	50 mL

Sterilize using 22 μm filtration filters, aliquot and store at −20 °C, and use for no longer than a year.

The aliquots can be frozen and defrozen several times.

2.3 N6-Benzyladenine (BAP, a Cytokinin)

Prepare a 100 μM stock solution in water.

To allow for the complete dissolution, add 1% v/v of NaOH 1 N.

Sterilize using 22 μm filtration filters, store aliquots at −20 °C and use for up to 6 months.

Once defrozen, the aliquots should be discarded after use.

2.4 Dyes to Stain Cell Contours

- Plasma membrane staining using FM4-64 (N-(3-triethylammoniumpropyl)-4-(6-(4-(diethylamino) phenyl) hexatrienyl) pyridinium dibromide).
  - Prepare a stock solution at 1 mg/mL in sterile water. Store aliquots at −20 °C and use for no longer than a year.
  - Once defrozen, the aliquots should be put back to −20 °C only once for maximal staining.
- Cell wall staining using propidium iodide (PI, 3, 8-diamino-5-[3-(diethylmethylammonio)propyl]-6-phenylphenanthridinium diiodide).
  - Use directly solution from Sigma-Aldrich available at 1 mg/mL (ref: P4864). Store in fridge at +4 °C.

2.5 Tools for Dissection

Tweezers: Size N°5 Roth (ref: LH79.1).

Petri dishes: 60 × 10 mm Falcon (ref: 353004).

Porous tape.

White and soft facial tissue with high absorption.

2.6 Microscope Equipment

- Stereo microscope.
- Confocal microscope upright (Zeiss LSM700) equipped with a long-distance objective X40 water immersion (e.g., we use a Zeiss W "N-Achroplan" 40×/0.75 M27 (ref: 420,967-9900), water, working distance 2.1 mm, without cover glass).
- Confocal microscope inverted (Zeiss LSM710) equipped with a long-distance objective X40 water immersion (e.g., we use a Zeiss W "N-Achroplan" 40×/1.0 M27 (ref: 421,462-9900), water, working distance 2.5 mm, without cover glass).

3 Methods

3.1 Preparation of the In Vitro Apex Culture Medium (ACM) Plates

1. Melt the half-MS medium.
2. Add vitamins (final concentration 1×) and BAP (final concentration 200 nM) under a laminar hood, sterile conditions (*see* **Note 1**). Pour the medium in Petri dishes.
3. Use directly or store at +4 °C for no longer than a month.

3.2 Plant Growth Conditions

1. Short-day conditions: 8 h light and 16 h night, 21 °C, hygrometry 55–60%.
2. Long-day conditions: 16 h light at 21 °C and 8 h night at 18 °C, hygrometry 55–60%.
3. To obtain healthy plant and facilitate dissection, plants are germinated in short-day condition and grown in these

Fig. 1 Illustration of the experimental procedure. (**A**) Optimal size of the plant for the dissection. (**B-E**) Several steps of the dissection procedure as described in the text. (**F**) Meristems on an ACM plate at the end of the dissection. (**G**) Setup for PI staining. (**H**) Wash after staining. (**I, J**) Setup for acquisition using upright (**I**) and inverted (**J**) microscope. (**K, L**) Visualization of fluorescence in a meristem on a Zeiss LSM710 inverted confocal microscope and Zen-associated software

conditions for 4 weeks. Plants are transferred for 2 weeks in long-day conditions to induce flowering and development of the inflorescence (optimal conditions, Fig. 1A).

4. Water the plants at least 2 h before dissection. It is possible to dissect meristems from plants grown in long days only or secondary and axillary meristems, but the smaller size of the stem makes the dissection more difficult.

3.3 Dissection

1. Dissect from 6 to 24 h before the first observation (*see*, **Note 2**).
2. First by naked eye, remove as much as possible the older flowers with the tweezers (Fig. 1B). Then using the stereo microscope, there are two alternative ways to continue with the dissection depending on the length of the inflorescence stem. Meristems with stems from 5 to 30 mm long can be dissected directly on the plant (Fig. 1C). Inflorescences longer than 30 mm can be cut and dissected on medium plate, by cutting at 1 cm for the top and transferring the segment of stem onto the medium plate (Fig. 1D).
3. Then remove old primordia one by one using tweezers; try to cut the petioles more closely to stem as possible. The meristem appears at the center as a green dark tissue in comparison with the primordia (Fig. 2) (*see* **Note 3**).
4. The last methods for very long stem and old plant: cut the stem at their base and place horizontally on a piece of craft paper and dissect by rolling the stem with the fingers (Fig. 1E), and remove all lateral organs progressively using tweezers (*see* **Note 4**).

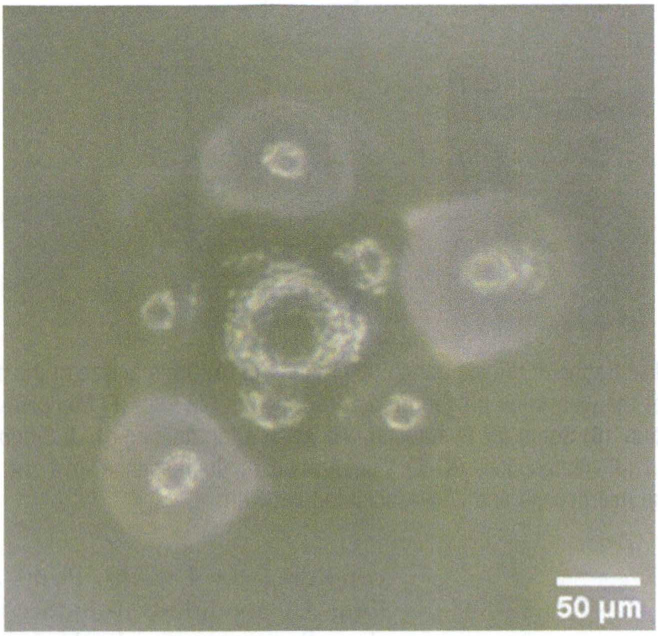

Fig. 2 Visualization under the stereo microscope of a dissected shoot apical meristem on the ACM medium. Note the darker color of the meristem at the center. Scale bar: 50 μm

5. After dissection, transfer the meristems to clean medium plates (*see* **Notes 5** and **6**) that will be used for microscope acquisition (Fig. 1F), taking care of aligning the stem with the gravity axis. Close the plates with porous tape and place the dishes back in the culture chamber or in a growth incubator with similar growth conditions.

3.4 Before Microscopy

1. Use a stereo microscope to check the dissected SAM as the sample might require additional dissection. Again, it is also important to check that the meristem is vertically aligned (Figs. 1F and 2).
2. Submerge the meristem with sterile water for 8 minutes by filling the dish (*see* **Note 7**).

3.5 Staining of the Cell Contours (Optional)

1. Stain meristems with PI (at 100 μM final concentration in water) for 5 min (*see* **Note 8**). In order to reduce staining volumes, we place a sterile 1000 μL tips (cutted about 2 cm from the top of the tip) around the dissected sample and add sufficient PI solution (about 150 μL) to fully cover the meristems (Fig. 1G). After incubation, rinse twice with sterile water (Fig. 1H) (*see* **Note 9**).
2. Alternatively SAMs can be stained with FM4-64. Add FM4-64 (at 300 μM final concentration in water) for 10 min by applying 10 μL (one drop) on top of the dissected meristem. The drop of FM4-64 solution should stay stably on the SAM for optimal staining. Rinse briefly twice with sterile water.

3.6 Acquisition

1. With an upright microscope (Fig. 1I), place the imaging dish onto the stage, and then lower the microscope lens until it touches the surface of the water (*see* **Note 10**), be careful to avoid air bubbles under the lens.
2. With an inverted microscope (Fig.1J), apply a drop of sterile water on the lens, hold the imaging dish upside down on the stage, and then raise the microscope lens until a water column is formed between the lens and medium (*see* **Notes 10** and **11**).
3. Using transmitted light, position the sample within field of the objective, and then focus on the top of the meristem using the Z control. It is also possible to use the epifluorescence illumination to find the meristem; however, this might lead to fluorescence bleaching of the sample and should be avoided if not absolutely necessary.
4. Adjust the confocal settings (laser power, gain, offset) accordingly to the signal. To avoid signal saturation, use the range indicator palette, and set the gain until there is no or few red

points in the image. It is recommended to use a minimal laser power to avoid photo-bleaching. It is also possible to open the pinhole up to 1.5–2.

5. We use the following settings: for VENUS (DR5, DII, and R2D2) laser 514 nm (5% power) and emission range 520–560 nm; for ntdTomato (R2D2) laser 561 nm (5% power) and emission range 580–620 nm; and for GFP (TCS) laser 488 (5% power) and emission range 500–550 nm. For PI and FM4-64, we used the 488 nm laser at 5% power, and adjust the emission range from 630 to 650 nm to avoid autofluorescence interference [15] (*see* **Note 12**).

3.7 After Imaging

1. When doing a time-course experiment (*see* **Notes 13** and **14**), it is important to remove the excess of water from the sample.
2. Under the stereo microscope, the meristem has to appear shiny and without any sign of water on it or on the medium before putting it back to grow. To remove all the water after imaging, use a clean absorbent facial tissue to absorb any drop on the medium and let the plate dry for 4–6 min before closing it.
3. When necessary, remove older organs after image acquisition not before to avoid stressing the meristem (*see* **Note 15**).

3.8 Analysis/Post Processing

1. For image processing and signal quantification, ImageJ or FiJi (http://imagej.net/Welcome or https://imagej.net/Fiji) offer multiple tools. It is out of our scope to describe all of them here, and we are just giving here a few hints on how to analyze the data. Maximal intensity Z-projections help to visualize the distribution of the different markers (Fig. 3).
2. Show projections obtained using Fiji, as in Fig. 3A; nuclear DR5 signal (LUT fire) and cell wall stained by PI (in yellow), in Fig. 3B; nuclear DII (LUT blue orange icb) and autofluorescence (in red), in Fig. 3C; reticulum endoplasmic TCS signal (LUT Green Fire Blue) and autofluorescence (in red), in Fig. 3D; R2D2 line with the DII signal (LUT Rainbow) and the constitutive expression (in green).
3. Further quantification and analysis of meristems can require tissue segmentation and 3D reconstruction. For this, MorphoGraphX MGX (http://www.mpipz.mpg.de/MorphoGraphX/P.) [16] or MARS-ALT (http://openalea.gforge.inria.fr/doc/vplants/vtissue/doc/_build/html/user/index.html [17] can be used.
4. We refer the reader to the corresponding website and publications for use of these approaches.

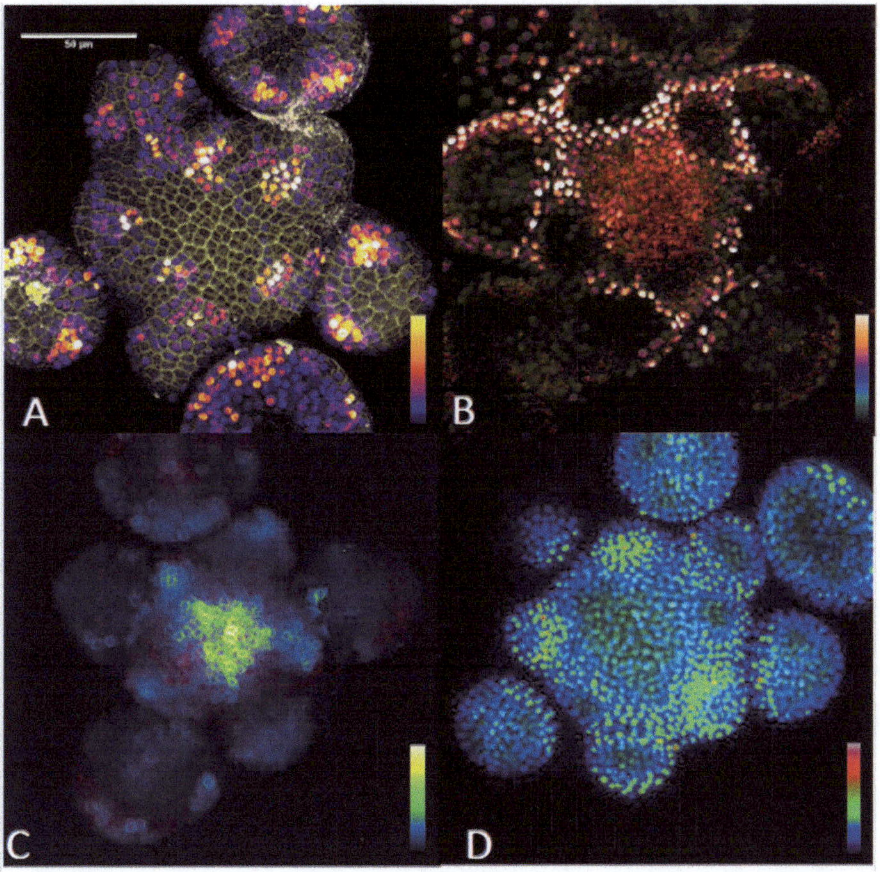

Fig. 3 Maximum projections performed using Fiji. (**A**) *DR5::3VENUS-N7* expression in LUT fire, cell wall stained by PI in yellow. (**B**) DII-VENUS in LUT blue orange icb, autofluorescence in red. (**C**) *TCS::erGFP* in LUT Green Fire Blue, autofluorescence in red. (**D**) R2D2; DII in LUT Rainbow, constitutive expression in green. Scale bar: 50 μm

4 Notes

1. Wait for medium temperature to reduce to 50–55 °C to avoid denaturation of vitamins and BAP.
2. Let the meristem recover from the dissection. Mechanical or wounding stresses are to be avoided as they can affect the fluorescence distribution of the reporters. DII is particularly sensitive and dynamics; the DII signal completely disappears within minutes after the dissection but re-appears progressively.
3. The dissection should be as rapid as possible to limit desiccation of the meristem.
4. The rolling method induces more mechanical stresses and a faster desiccation of the sample but can be very useful when the plants have grown more rapidly than expected.

5. Let the plate be at room temperature at least 30 min before using to avoid cold stress to the meristem.
6. Dissect a double number of meristems than the number you expect to image and analyze because some of them will not survive over a long time course.
7. It is crucial to immerge the sample in water few minutes before imaging to avoid sample movement during acquisition due to water absorption and swelling of the gel.
8. PI staining reveals damage, because when the tissue is damaged, the PI will penetrate inside and the signal will be strong and/or in the nucleus rather than well localized at the cell wall.
9. The cutted 1000 μL tips can then be reused. They need then to be kept in ethanol 70% to avoid contamination.
10. Use a water immersion objective to avoid desiccation; if dry objective is used, the imaging time should be less than 1 min.
11. During long imaging time, regularly check the water column and add water if necessary.
12. When imaging meristems with different fluorophores, we recommend sequential (multitrack) acquisition to avoid cross-channel signal contamination.
13. Bacteria or fungus contamination will kill the meristem. To avoid this, use a new bottle of sterile water at each imaging time. The objective has also to be cleaned before imaging with 70% ethanol.
14. For long time courses (more than 2 days), the ACM medium has to be changed (every other day) to avoid contamination but also be certain that the BAP concentration is stable.
15. A way to know the meristem is still alive and in good condition is that you should detect new organ initiation at around one new organ per 12 h.

References

1. Petrasek J, Friml J (2009) Auxin transport routes in plant development. Development 136:2675–2688
2. Pfeiffer A, Wenzl C, Lohmann JU (2017) Beyond flexibility: controlling stem cells in an ever-changing environment. Curr Opin Plant Biol 35:117–123
3. Besnard F, Rafahi Y, Morin V, Marteaux B, Brunoud G, Chambrier P, Frédérique R, Mirabet V, Legrand J, Lainé S, Thévenon E, Farcote E, Cellier C, Das P, Bishopp A, Dumas R, Parcy F, Helariutta Y, Boudaoud A, Godin C, Traas J, Vernoux T (2013) Cytokinin signalling inhibitory fields provide robustness to phyllotaxis. Nature 550:417–421
4. Brunoud G, Wells DM, Oliva M, Larrieu A, Mirabet V, Beeckman T, Kepinski S, Traas J, Bennett MJ, Vernoux T (2012) A novel sensor to map auxin response and distribution at high spatio-temporal resolution. Nature 482:103–106
5. Liao C, Smet W, Brunoud G, Yoshida S, Vernoux T, Weijers D (2012) Reporters for sensitive and quantitative measurement of auxin response. Nat Methods 12:207–210
6. Ulmasov T, Murfett J, Hagen G, Guilfoyle TJ (1997) Aux/IAA proteins repress expression of

reporter genes containing natural and highly active synthetic auxin response elements. Plant Cell 9:1963–1971

7. Sabatini S, Beis D, Wolkenfelt H, Murfett J, Guilfoyle T, Malamy J, Benfey P, Leyser O, Bechtold N, Weisbeek P, Scheres B (1999) An auxin dependent distal organizer of pattern and polarity in the Arabidopsis root. Cell 99:463–472
8. Muller B, Sheen J (2008) Cytokinin and auxin interaction in root stem-cell specification during early embryogenesis. Nature 453:1094–1097
9. Zürcher E, Tavor-Deslex D, Lituiev D, Enkerli K, Tarr PT, Müller B (2013) A robust and sensitive synthetic sensor to monitor the transcriptional output of the cytokinin signalling network in plant. Plant Physiol 161:10066–11075
10. Benkova E, Michniewicz M, Sauer M, Teichmann T, Seifertova D, Jürgens G, Friml J (2003) Local, efflux-dependent auxin gradients as a common module for plant organ formation. Cell 115:591–602
11. Heisler MG, Ohno C, Das P, Sieber P, Reddy GV, Long JA, Meyerowitz E (2005) Patterns of auxin transport and gene expression during primordium development revealed by live imaging of the Arabidopsis inflorescence meristem. Curr Biol 15:1899–1911
12. Vernoux T, Brunoud G, Farcot E, Morin V, Van den Daele H, Legrand J, Oliva M, Das P, Larrieu A, Wells D, Guédon Y, Armitage L, Picard F, Guyomarc'h S, Cellier C, Parry G, Koumproglou R, Doonan JH, Estelle M, Godin C, Kepinski S, Benette M, De Veylder L, Traas J (2011) The auxin signalling network translate dynamic input into robust patterning at the shoot apex. Mol Syst Biol 7:508
13. Adibi M, Yoshida S, Weijers D, Fleck C (2016) Centering the organizing center in the Arabidopsis thaliana shoot apical meristem by combination of cytokinin signalling and self-organization. PLoS One 11:e0147830
14. Liao C, Smet W, Brunoud G, Yoshida S, Vernoux T, Weijers D (2015) Corrigendum: reporters for sensitive and quantitative measurement of auxin response. Nat Methods 12:1098
15. Chudakov DM, Matz MV, Lukyanov S, Lukyanov KA (2010) Fluorescent proteins and their applications in imaging living cells and tissues. Physiol Rev 90:1163–2010
16. Barbier de Reuille P, Robinson S, Smith RS (2015) Quantification cell shape and gene expression in the shoot apical meristem using MorphographX. Plant cell morphogenesis: methods and protocols. Methods Mol Biol 1080:121–134
17. Fernandez R, Das P, Mirabet V, Moscardi E, Traas J, Verdeil J-L, Malanchain G, Godin C (2010) Imaging plant growth in 4D: robust tissue reconstruction and lineaging at cell resolution. Nat Methods 7:547–553

Chapter 10

Analysis of Cell Division Frequency in the Root Apical Meristem of Lycophytes, Non-seed Vascular Plants, Using EdU Labeling

Rieko Fujinami

Abstract

The organization of the root apical meristem (RAM) provides insights into the evolution of roots in vascular plants. The RAM of seed plants has a quiescent center (QC), in which the cells divide infrequently and function to maintain neighboring stem cells. However, the existence of a QC and the mechanisms of RAM maintenance in non-seed plants are poorly understood. We analyzed the RAM organization of lycophytes focusing on cell division activity using the EdU labeling method and showed that the RAM of *Lycopodium* species has a region with a very low cell division frequency, which was named the QC-like region. Here, we describe an in situ EdU labeling method for the RAM of growing roots in nature.

Key words Cell division, EdU, Histology, Lycophyte, Root apical meristem, Quiescent center

1 Introduction

The root apical meristem (RAM) of seed plants possesses a population of infrequently dividing cells in its center. This population of cells was first detected by the [3H]thymidine labeling method and was named the quiescent center (QC) by Clowes [1, 2]. Recent molecular genetic analyses further clarified that the QC is surrounded by stem cells (initial cells) that act as a reservoir of undifferentiated cells [3–5] and that the QC functions to maintain the stem cell niche [6, 7].

Following Clowes's proposal [1, 2], studies of cell proliferation in the root apical meristem were performed using [3H]thymidine labeling [8, 9]. Later, another detection method, BrdU labeling, was widely adopted because it was easier than autoradiography. In the BrdU labeling method, dividing cells were detected by fluorescence immunostaining of DNA that had incorporated a uridine analog, 5-bromo-2′-deoxyuridine (BrdU) [10, 11]. However, the BrdU labeling method was superseded by 5-ethynly-

Muhammad Naseem and Thomas Dandekar (eds.), *Plant Stem Cells: Methods and Protocols*, Methods in Molecular Biology, vol. 2094, https://doi.org/10.1007/978-1-0716-0183-9_10, © Springer Science+Business Media, LLC, part of Springer Nature 2020

2′deoxyuridine (EdU) labeling because the latter has several advantages. EdU is a thymidine analog, with replacement of the 5′ methyl group in the pyrimidine ring with an alkyne. To visualize incorporation of EdU, the alkyne is reacted with a fluorescent azide in a Cu (I)-catalyzed reaction, rather than the use of fluorescence immunostaining of antibodies as in BrdU labeling [12]. The fluorescent azide is small in size and can rapidly penetrate tissues and organs, allowing efficient detection of the incorporated EdU [12–15]. Thus, detection of EdU is highly sensitive and much faster than that of BrdU. At present, the EdU labeling method combined with molecular genetic analyses has yielded good results regarding the mechanisms of RAM maintenance and development in angiosperm roots, especially *Arabidopsis* [16, 17].

The QC was believed to be absent in non-seed plants. However, we recently demonstrated the occurrence of a region of very low cell division frequency, similar to the QC, in the RAM of a lycophyte, *Lycopodium clavatum*, by analysis of the frequency and pattern of mitotic divisions using the EdU labeling method [18]. This is also the first report regarding the analysis of cell division in growing roots in nature. Furthermore, lycophytes are the earliest diverging lineage of vascular plants and represent a key taxon to investigate the evolution of three organs, i.e., the root, stem, and leaf, in vascular plants. EdU labeling analysis should be performed across other lycophyte taxa. Here, we describe the EdU labeling method of growing roots in the field and calculation of the cell division frequency in the RAM.

2 Materials

2.1 EdU Labeling of Lycopodium RAM

1. EdU: EdU is available in Click-iT EdU Imaging Kits (Molecular Probes Inc., Eugene, OR, USA).
2. Eppendorf tubes or tubes of suitable sizes for growing roots.
3. 50-mL centrifuge tubes.
4. Vinyl tape.

2.2 Embedding and Sectioning for Lycopodium RAM

1. Fixative: 4.0% PFA in 0.1 M phosphate buffer (pH 7.2).
2. Microscope slides.
3. Cover glasses.
4. EtOH.
5. Molds to embed materials.
6. Technovit 7100 (Heraeus Kulzer, Hanau, Germany).
7. Microtome and blades.

Table 1
Click-iT reaction buffer for slides

	Number of slides		
Reaction components	**1**	**5**	**10**
1X Click-iT® reaction buffer	430 μL	2.2 mL	4.3 mL
$CuSO_4$	20 μL	100 μL	200 μL
Fluorescent azide	1.2 μL	6 μL	12.5 μL
Reaction buffer additive	50 μL	250 μL	500 μL
Total volume	500 μL	2.5 mL	5 mL

For the reaction to proceed optimally, the components are added in the order listed in the table, and the buffer is used immediately (within 15 min)

2.3 EdU Detection of *Lycopodium* RAM

1. PBS.
2. Permeabilization buffer, 0.5% Triton X-100 in PBS.
3. Wash buffer, 3% BSA in PBS.
4. EdU detection reagents, Click-iT EdU Imaging Kits (Molecular Probes Inc.), which will be described in detail in this protocol (Table 1).
5. DAPI for DNA counterstaining (stock solution 1 mg/mL).
6. Parafilm®.
7. Humid chamber.
8. Fluorescent microscope and phase-contrast microscope.
9. Light microscope.

3 Methods

3.1 EdU Labeling of *Lycopodium* RAM

1. Prepare a solution of EdU in Eppendorf tubes or other tubes of a suitable size. First, make a stock solution of EdU following the protocols supplied with the Click-iT EdU Imaging Kit, and dilute with water just prior to use. The recommended concentration for the EdU stock solution is 10 μM.
2. In the field, immerse the root tip of the plant (here *Lycopodium clavatum*) in the EdU solution in the tube (Fig. 1, *see* **Note 1**).
3. Label RAMs with EdU for 24 h (*see* **Note 2**).

3.2 Fixation and Sectioning of Materials

1. After 24 h of immersion, remove the root from the tube, and cut the root tip from the body of the plant.
2. Place the root tips in 50-mL centrifuge tubes with fixative (*see* **Note 3**).
3. Incubate the roots in fixative solution overnight at 4 °C.

Fig. 1 Examination of EdU labeling in a root of *Lycopodium clavatum* in nature. The root was placed into a tube containing EdU solution and held in place with vinyl tape (arrow). The arrowhead indicates the root apex of *Lycopodium clavatum*. *R* root, *S* stem

4. Remove the fixative solution from the centrifuge tube, and wash the roots with three changes of 0.1 M phosphate buffer solution (pH 7.2) in the centrifuge tubes.
5. Replace the buffer with 10% ethanol and dehydrate for 30 min. Then, dehydrate the root tips with 20%, 30%, and 40% ethanol for 30 min, followed successively by 50%, 60%, 70%, 80%, 90%, 95%, and 100% (absolute) ethanol for 1 h each.
6. Replace the absolute ethanol with Technovit 7100 resin for 2–4 days (*see* **Note 4**).
7. Place root tips in molds containing Technovit 7100 at room temperature (*see* **Note 5**), and leave for 2–4 h.
8. Remove the solidified resin-embedded samples from the molds, and trim off resin surrounding the root tips as much as possible using a razor blade or scalpel (*see* **Note 6**).
9. Set a glass knife in the microtome, and section the tissue. We usually cut sections at a thickness of 2 μm.
10. Place the sections on a drop of distilled water on a microscope slide to expand shrunken tissues.
11. Evaporate the remaining water from the microscope slides using a hot plate at 50 °C.

3.3 EdU Detection in Sectioned RAMs

1. Prepare Click-iT reaction buffer according to Table 1. Use the Click-iT reaction buffer within 15 min after preparation (*see* **Note 7**).
2. Transfer microscope slides with tissue sections to a humid chamber.
3. Add 1 mL of 0.5% Triton®X-100 in PBS to each slide, and then incubate at room temperature for 10 min.
4. Remove the 0.5% Triton®X-100 in PBS, and wash the sections twice with 1 mL PBS.
5. Add 500 μL of Click-iT reaction buffer to each slide, and cover with a Parafilm®. Place these slides in a closed humid chamber, and incubate at room temperature for 45 min, shielded from the light (*see* **Note 8**).
6. Remove the Parafilm® and wash slides three times with PBS for 20 min each time (*see* **Note 9**).
7. Prepare DAPI working solution by diluting stock solution 1:1000 in PBS.
8. Add 500 μL of DAPI working solution to the slide and cover with Parafilm®. Place in a closed humid chamber and incubate at room temperature for 5 min.
9. Remove the Parafilm® and wash the slide once with PBS for 10 min.
10. Visualize the tissue with a fluorescence microscope and a phase-contrast microscope (*see* **Note 10**).

3.4 Fluorescent Image Capture of RAM Sections

1. After capturing EdU images, DAPI images, and phase-contrast images with the fluorescence microscope and phase-contrast microscope (Fig. 2), remove the cover glasses from the slides, and dry at 50 °C (*see* **Note 11**).
2. The sections on the slides are stained with Sharman's staining solution for histological observation (*see* **Note 12**).
3. Take histological images under a light microscope.

3.5 Analysis of the Cell Division Frequency of *Lycopodium* RAM

1. To determine the spatial distribution of mitotic cells in the *Lycopodium* RAM, a combined image is obtained by superposition of the EdU-treated images of five serial sections using Adobe Photoshop CS software (Adobe Systems Inc., San Jose, CA) (Fig. 3).
2. To determine the percentage of cells undergoing mitosis per total cells in the RAM, an EdU image, DAPI image, and phase-contrast image of the same median longitudinal section are superimposed (Fig. 3).

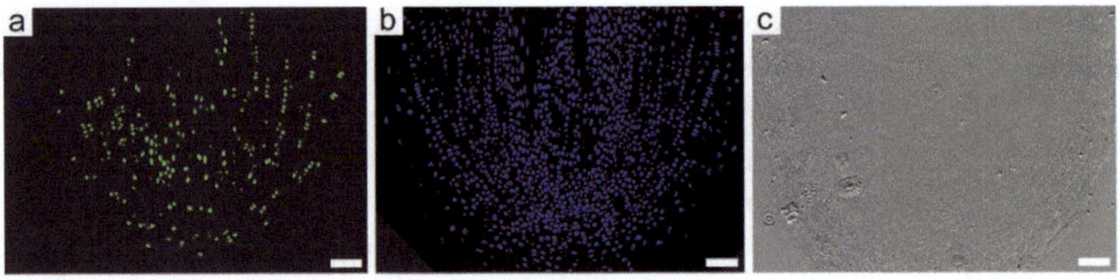

Fig. 2 EdU image (**a**), DAPI image (**b**), and phase-contrast image (**c**) of *Lycopodium clavatum* RAM. Bars, 100 μm

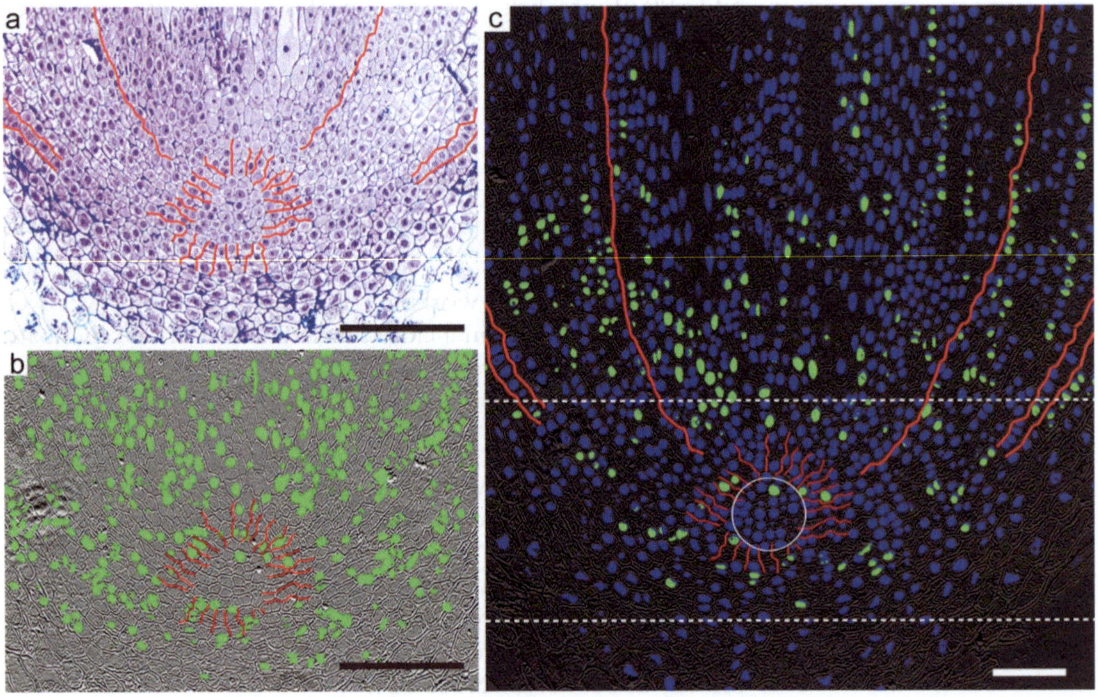

Fig. 3 *Lycopodium clavatum* root apical meristem (RAM) visualized using histochemical staining (**a**) and EdU staining (**b**, **c**) by a modification of the method of Fujinami et al. [18]. The circle indicates the common initial zone of the *L. clavatum* RAM. (**a**) Median longitudinal section of *L. clavatum* RAM. Raced lines in the center of the RAM indicate the layers of the initial cells. (**b**) Superposition of EdU images of five serial sections and a phase-contrast image of a median longitudinal section. (**c**) Image of EdU-labeled cells superimposed over DAPI and phase-contrast images of a median longitudinal section. The circle indicates the QC-like area of the *L. clavatum* RAM. The area between the two lines indicates the area of measurement of the cell division frequency. Bars, 100 μm

3. To calculate the percentage of mitotic cells, the numbers of EdU-labeled and DAPI-stained nuclei are counted separately in the RAM area (*see* **Note 13**).
4. The percentage of EdU-labeled nuclei is calculated as follows: number of green cells divided by the number of blue + green nuclei.

4 Notes

1. We anchored the tube containing EdU buffer to the plant body using vinyl tape (Fig. 1). To prevent it from being damaged, the root apex must not touch the bottom of the tube.
2. The time of EdU exposure depends on the cell growth rate of the tissues examined. Based on preliminary experiments, we used an exposure time of 24 h for the *Lycopodium* RAM in the EdU solution. Additionally, the concentration of EdU is also dependent on the plant tissue examined. Among 5, 10, and 20 μM EdU, we obtained the best results with 10 μM EdU.
3. As the fixative solution, we successfully used 4.0% paraformaldehyde in 0.1 M phosphate buffer (pH 7.2) with vacuum infiltration. We also obtained good results with PFA as the fixative solution. Depending on the samples, formaldehyde may also be used. The fixative times listed in this protocol are specifically for PFA.
4. Replace from ethanol to Technovit 7100 gradually as follows:

   Technovit 7100:100% EtOH = 1:2, 2 h, 4 °C.

   Technovit 7100:100% EtOH = 2:2, 2 h, 4 °C.

   Technovit 7100:100% EtOH = 2:1, 12 h, 4 °C.

   Technovit 7100, 24 h, 4 °C × 2.
5. For polymerization, 1 mL of hardener is added to 15 mL of Technovit 7100 and mixed well. The mixed Technovit 7100 solution is quickly put into the molds, and the samples are then placed into the solution.
6. We recommend examining the samples by EdU labeling within 6 months after embedding in Technovit.
7. To visualize dividing cells in mitosis in a RAM with EdU labeling, we used Alexa 488, which requires a different fluorescence wavelength from DAPI.
8. As a humid chamber, a closed plastic container with a moist paper towel at the bottom is used. The container must be large enough to place several microscope slides on the moist paper.
9. To keep the background of EdU detection at a minimum, it is necessary to wash the slides with PBS more than three times.
10. On microscope slides, we marked five serial sections using a marker pen to facilitate capture of EdU, DAPI, and phase-contrast images of the same sections.
11. The cover glass should be removed carefully from the microscope slide. To facilitate removal, we add a small amount of distilled water between the cover glass and the microscope slide. This allows the floating cover glass to be removed.

Microscope slides with sections are then dried by evaporation using a hot plate.

12. Slides with tissue sections are stained with a modified Sharman's stain [19], which includes 5–6 min in 0.2% toluidine blue in 0.2% sodium tetraborate between the usual safranin O and orange G-tannic acid staining steps.
13. The EdU-labeled and DAPI-stained nuclei are counted in an image of the median longitudinal section of the RAM using the Count tool in Adobe Photoshop software.

Acknowledgment

The author is grateful to Prof. Ryoko Imaichi for helpful comments on the manuscript. The images shown in Fig. 3 were published in New Phytologist. This work was supported by a Grant-in-Aid (KAKENHI) for Scientific Research (no. 25870088) from the Japan Society for the Promotion of Science (JSPS) and a Plant Research Grant from the New Technology Development Foundation (Japan).

References

1. Clowes FAL (1956) Localization of nucleic acid synthesis in root meristems. J Exp Bot 7:307–312
2. Clowes FAL (1956) Nucleic acid in root apical meristems of *Zea*. New Phytol 55:29–34
3. Jiang K, Feldman L (2003) Root meristem establishment and maintenance: the role of Auxin. J Plant Growth Regul 21:432–440
4. Sabatini S, Heidstra R, Wildwater M, Scheres B (2003) SCARECROW is involved in positioning the stem cell niche in the *Arabidopsis* root meristem. Genes Dev 17:354–358
5. Sarkar AK, Luijten M, Miyashima S, Lenhard M, Hashimoto T, Nakajima K, Scheres B, Heidstra R, Laux T (2007) Conserved factors regulate signaling in *Arabidopsis thaliana* shoot and root stem cell organizers. Nature 446:811–814
6. Ortega-Martínez O, Pernas M, Carol RJ, Dolan L (2007) Ethylene modulates stem cell division in the *Arabidopsis thaliana* root. Science 317:507–519
7. Heyman J, Cools T, Vandenbussche F, Heyndrickx KS, Leene JV, Vercauteren I, Vanderauwera S, Vandepoele K, Jaeger GD, Straeten DVD, Veylder LD (2013) ERF115 controls root quiescent center cell division and stem cell replenishment. Science 342:860–863
8. Clowes FAL (1961) Apical meristems. Blackwell, Oxford, UK
9. Dolan L, Janmaat K, Willemsen V, Linstead P, Poethig S, Roberts K, Scheres B (1993) Cellular organisation of the *Arabidopsis thaliana* root. Development 119:71–84
10. Fujie M, Kuroiwa H, Suzuki T, Kawano S, Kuroiwa T (1993) Organelle DNA synthesis in the quiescent centre of *Arabidopsis thaliana* (Col.). J Exp Bot 44:689–693
11. Ogawa A, Kitamichi K, Toyofuku K, Kawashima C (2006) Quantitative analysis of cell division and cell death in seminal root of Rye under salt stress. Plant Prod Sci 9:56–64
12. Salic A, Mitchison TJ (2008) A chemical method for fast and sensitive detection of DNA synthesis in vivo. Proc Natl Acad Sci U S A 105:2415–2420
13. Kotogány E, Dudits D, Horváth GV, Ayaydin F (2010) A rapid and robust assay for detection of S-phase cell cycle progression in plant cells and tissues by using ethynyl deoxyuridine. Plant Methods 6:5
14. Hayashi K, Hasegawa J, Matsunaga S (2013) The boundary of the meristematic and elongation zones in roots: endoreduplication precedes rapid cell expansion. Sci Rep 3:2723. https://doi.org/10.1038/srep02723

15. Kazda A, Akimcheva S, Watson JM, Riha K (2016) Cell proliferation analysis using EdU labeling in whole plant and histological samples of Arabidopsis. In: Caillaud M-C (ed) Plant cell division: methods and protocols, methods molecular biology. Springer, New York
16. Vanstraelen M, Baloban M, Da Ines O, Cultrone A, Lammens T, Boudolf V, Brown SC, De Veylder L, Mergaert P, Kondorosi E (2009) APC/C^{CCS52A} complexes control meristem maintenance in the *Arabidopsis* root. Proc Natl Acad Sci U S A 106:11806–11811
17. Perilli S, Perez-Perez JM, Di Mambro R, Peris CL, Diaz-Trivino S, Del Bianco M, Pierdonati E, Moubayidin L, Cruz-Ramirez A, Costantino P, Scheres B, Sabatini S (2013) RETINOBLASTOMA-RELATED protein stimulates cell differentiation in the *Arabidopsis* root meristem by interacting with cytokinin signaling. Plant Cell 25:4469–4478
18. Fujinami R, Yamada T, Nakajima A, Takagi S, Idogawa A, Kawakami E, Tsutsumi M, Imaichi R (2017) Root apical meristem diversity in extant lycophytes and implications for root origins. New Phytol 215:1210–1220
19. Jernstedt JA, Cutter EG, Gifford EM, Lu P (1992) Angle meristem origin and development in *Selaginella martensii*. Ann Bot 69:351–363

Chapter 11

Osmotic Treatment for Quantifying Cell Wall Elasticity in the Sepal of *Arabidopsis thaliana*

Aleksandra Sapala and Richard S. Smith

Abstract

Elastic properties of the cell wall play a key role in regulating plant growth and morphogenesis; however, measuring them in vivo remains a challenge. Although several new methods have recently become available, they all have substantial drawbacks. Here we describe a detailed protocol for osmotic treatments, which is based on the idea of releasing the turgor pressure within the cell and measuring the resulting deformation. When placed in hyperosmotic solution, cells lose water via osmosis and shrink. Confocal images of the tissue, taken before and after this treatment, are quantified using high-resolution surface projections in MorphoGraphX. The cell shrinkage observed can then be used to estimate cell wall elasticity. This allows qualitative comparisons of cell wall properties within organs or between genotypes and can be combined with mechanical simulations to give quantitative estimates of the cells' Young's moduli. We use the abaxial sepal of *Arabidopsis thaliana* as an easily accessible model system to present our approach, but it can potentially be used on many other plant organs. The main challenges of this technique are choosing the optimal concentration of the hyperosmotic solution and producing high-quality confocal images (with cell walls visualized) good enough for segmentation in MorphoGraphX.

Key words Plasmolysis, Cell wall, Shrinkage, Biomechanics, Sepal

1 Introduction

Plant cells are encased by cell walls that need to be extremely durable in order to withstand the high turgor pressure within. This pressure is the effect of higher osmotic potential of the protoplast compared to extracellular space and typically results in pressure of around 2–10 bar in *Arabidopsis* tissue [1, 2]. Although the cell wall must be rigid enough to contain turgor pressure, it also needs to relax in a controlled fashion to allow cell and tissue growth. Studies have shown that growing cells have softer cell walls than nongrowing cells [3–6]. To quantify cell wall stiffness changes in the context of the genes controlling morphogenesis, new methods such as Brillouin microscopy [7], atomic force microscopy (AFM) [3, 8–11], and other microindentation systems

Muhammad Naseem and Thomas Dandekar (eds.), *Plant Stem Cells: Methods and Protocols*, Methods in Molecular Biology, vol. 2094, https://doi.org/10.1007/978-1-0716-0183-9_11, © Springer Science+Business Media, LLC, part of Springer Nature 2020

have been applied to plants [12–15]. However, all of these methods have a limited ability to measure cell wall stiffness in the in-plane direction of the cell wall, the direction that is presumed to be the most relevant for growth and morphogenesis. Although direct methods with extensometer are the most reliable in this respect [16], they can only be applied in-planta in very limited situations [17–19].

Osmotic treatments offer the possibility to measure in-plane cell wall stiffness in living plant cells [4, 6, 15, 20] in a relatively direct manner. Like an extensometer, osmotic treatments are based on changing the stress on the cell wall and recording the deformation that occurs as a result. Cells are imaged with a confocal laser scanning microscope, and the change in their surface area is quantified using the 3D image processing software MorphoGraphX [4, 21]. The change in stress is accomplished by manipulating the cell's turgor pressure through osmosis. The turgor pressure in plant cells results from the difference in osmotic potential between the cell cytosol and the extracellular space. Since the cell cytosol has a higher osmolarity than extracellular space, the cell will take up water through osmosis until the physical pressure balances the difference in osmotic potential. Here we increase the osmotic potential of the extracellular space by immersing the sample in a solution with osmolarity equal to or slightly higher than the cytosol. This causes the cells to deflate, a process called plasmolysis, and can be used to estimate the turgor pressure of the cells (for a review, *see* [22]).

The amount by which the cell changes its size (deformation) gives an estimate of the elasticity of the cell walls. Young's modulus (E) is a measure of cell wall stiffness and is defined as the ratio of mechanical stress over strain (deformation of a material induced by the stress divided by the material's initial length). Therefore, the larger the shrinkage of the cell surface observed in the experiment, the lower the Young's modulus and the softer the material.

Inferring the exact E values is far less trivial, as strains can be different along different directions and can depend on other material parameters such as the Poisson's ratio. Hence, to achieve quantitative results, a reverse engineering approach based on mechanical simulations, most often with the finite element method (FEM), is required [20, 23]. Nevertheless, comparisons of cell wall elasticity within an organ or between different genotypes of the same species in the same tissue type can give a qualitative measure of differences in stiffness [6]. The more a cell shrinks upon osmotic treatment, the more elastic it is.

Here we present the use of this method on the *Arabidopsis thaliana* sepal, which is a convenient system to answer basic questions about young, growing cells. As the outermost floral organ, the abaxial sepal is easily accessible for dissection and confocal imaging. The epidermis of the sepal is comprised of large, elongated cells called giant cells randomly interspersed between smaller polygonal cells called small cells [24]. Its growth patterns are described as a wave of anisotropic growth moving from the top to the bottom of the organ throughout its development [25], and we have used this method to quantify sepal cell wall stiffness, which is lower in the fast-growing zones than in the slow-growing zones [6].

2 Materials

1. Flowering plants of *Arabidopsis thaliana* soil-grown in long-day conditions. Apart from your genotype of interest and wild-type control, use plants with a fluorescent marker in the plasma membrane in order to confirm that plasmolysis is taking place (*see* **Note 1**). We have used the *pUBQ10::myrYFP* line used in [25].
2. Binocular and forceps.
3. A medical needle, with a small hook (scratch the needle along a hard surface to get the hook).
4. 1.5 mL or smaller Eppendorf tube.
5. 0.1% solution of plant protective medium in water (referred to as PPM water).
6. 0.1% solution of propidium iodide in water.
7. Plastic Petri dishes half-filled with ½ MS agar medium containing PPM, prepared in sterile conditions.
8. 0.4 M solution of NaCl in water.
9. Confocal laser scanning microscope with 20× or 40× water immersion lens.
10. A computer with an NVIDIA 3D graphics card and Linux (Mint/Debian/Ubuntu) to install MorphoGraphX (www.morphographx.org).

3 Methods

3.1 Tissue Dissection

1. One day before the experiment, water the plants.
2. Take a flowering *A. thaliana* plant. Cut off the stem 3–4 cm below the inflorescence meristem. Put the removed part under a dissecting microscope.
3. Using a syringe needle, cut off the largest flowers from the proximity of the inflorescence meristem.
4. Pick one flower of about 200–300 μm in width. We recommend flowers that are at stage 9 of development as described in [26]. Remove all the other flowers from the stem.

3.2 Osmotic Treatment and Image Acquisition

1. Fill an Eppendorf tube with 0.1% PPM water. Insert the dissected stem, taking care that the flower is under water. Incubate for 1 h.
2. Transfer the sample into another Eppendorf tube filled with 0.1% PI solution. Incubate for 15 min.
3. Put the sample into the ½ MS solid medium (stem in the medium, flower sticking out, with the abaxial flower as parallel to the medium surface as possible).
4. Determine the optimal concentration of osmotic agent for your study system (*see* **Note 1**). It should be enough to plasmolyze the cells but not so high as to stress them.
5. Fill the Petri dish with water or the osmotic agent solution and take the first image with the confocal microscope. This will be time point 0 (T0). *See* **Note 2** for an explanation of which liquid to use for imaging T0 and **Note 3** for microscope settings.
6. Incubate for 30 min.
7. Take another confocal image with the same settings as in pt. 5. This will be time point 1 (T1). *See* **Notes 3** and **4**.

3.3 Image Quantification with MorphoGraphX

In order to track the relatively small deformations from cell shrinkage, a precise way to quantify changes in cell size and shape is required. Here we present the basic protocol for image processing in MorphoGraphX, a software specifically designed for this purpose [4]. We focus on the points critical for successful analysis of osmotic treatment data; however, for a more extensive description of functions and workflows available in MorphoGraphX, we refer the reader to [21] and the MorphoGraphX user manual (http://www. MorphoGraphX.org/help).

1. Load the T0 stack into MorphoGraphX as a tiff file. Make sure that the file contains image from one microscope detection channel only. *See* **Note 5** for instructions on how to handle the stack within MorphoGraphX.
2. If your image was not taken at 16-bit resolution, use the "Stack/Filters/Brighten Darken" process with amount set to 16 for 12-bit images or to 256 for 8-bit images. Note that images loaded from 8-bit tiffs might be converted automatically.
3. Blur the stack using the "Stack/Filters/Gaussian Blur Stack" process.
4. Extract the shape of your sample using the "Stack/Morphology/Edge Detect" process (*see* **Note 6**).

5. Create a mesh using the "Mesh/Creation/Marching Cubes Surface" process. Then trim off the bottom of the mesh as well as parts you do not need (*see* **Note 7**).
6. Smooth the mesh using the "Mesh/Structure/Smooth Mesh" process, and subdivide it using the "Mesh/Structure/Subdivide" process (*see* **Note 8**). Keep in mind to save your mesh after each step (*see* **Note 9**).
7. Project signal from the top 2–6 μm of your confocal stack onto the mesh ("Mesh/Signal/Project Signal"). Now you should clearly see cell outlines on the mesh. Adjust the depth of projected signal in each sample to get the best result.
8. Using the "Add New Seed" tool (working only with Alt pressed on your keyboard), put one seed in each cell you want to segment (*see* **Note 10**).
9. Draw a "border seed" around all cells already containing a seed.
10. Run the "Mesh/Signal/Smooth Mesh Signal" process.
11. Segment your mesh using the "Mesh/Segmentation/Watershed Segmentation" process.
12. Switch between the "Cells/Labels" and "Vtx/Signal" in the Mesh menu (Fig. 1) to locate under- or over-segmented cells. Correct segmentation mistakes (*see* **Note 11**).
13. Run the "Mesh/Cell Mesh/Fix Corners" process in order to cover all unseeded vertices.
14. Now, you have completed the segmentation (Fig. 2b). Save your mesh, clearly indicating the time point.
15. Load the next stack (T1) and repeat **steps 2–14**. Caution: the post-osmotic-treatment confocal picture will be of lower quality (Fig. 2a). Please *see* **Notes 12** and **13** for guidelines on how to deal with this.
16. Match corresponding cells (or cell groups) from the two time points to calculate cell shrinkage (Fig. 2c). Start by loading segmented T0 and T1 meshes into the Stack 1 and Stack 2 channels, respectively.
17. In the "View" tab, set the Ctrl key interaction to "Stack 1." This way, you can move it independently of Stack 2 when you press Ctrl on your keyboard. Then, superimpose the two stacks in a way that at least a few corresponding cells clearly match. *See* **Note 14**.
18. Deactivate the surface in Stack 1. Tick the "Parents" box in the "Main" tab for Stack 2. With Stack 2 active, use the "Grab label from another surface" tool and click on cells in Stack 2 through cells in Stack 1. The cells will fill with the corresponding label from Stack 1, called "parent label" as this process was originally

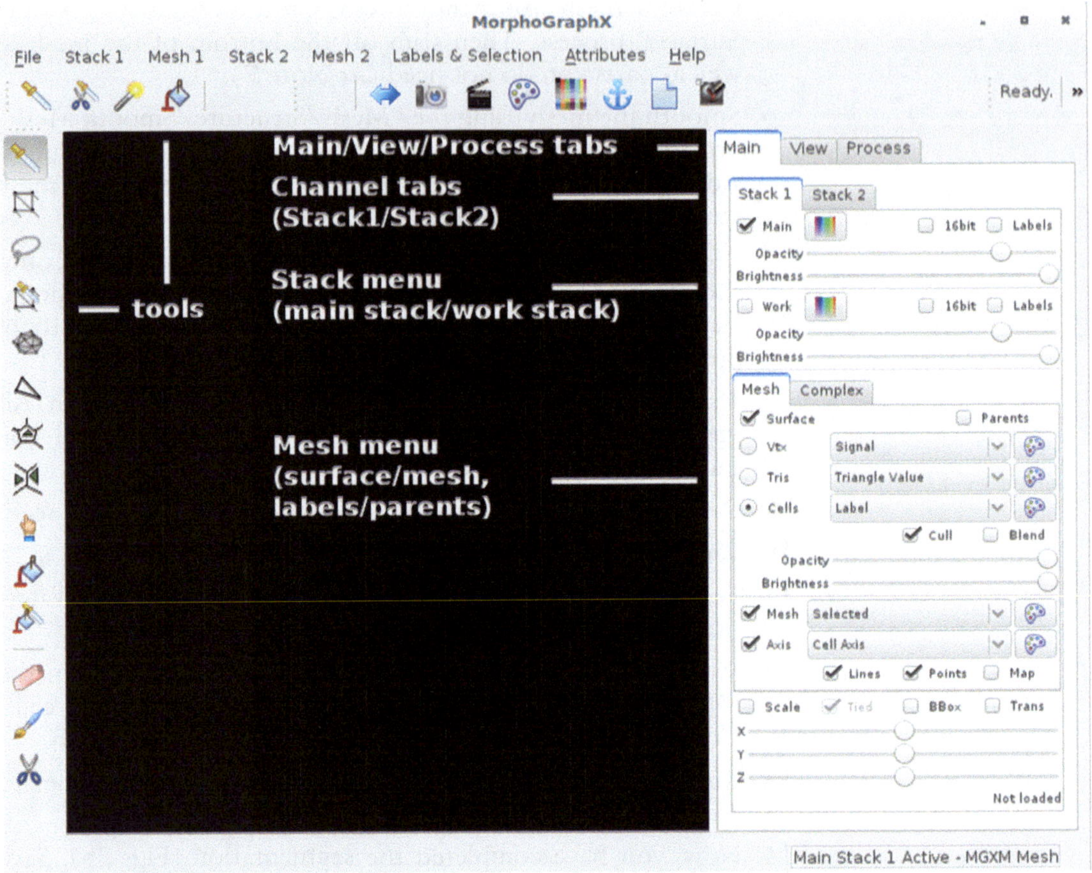

Fig. 1 The MorphoGraphX user interface. There are two channels called Stack 1 and Stack 2. A confocal stack can be loaded into both, so that two separate stacks can be worked on independently. When running any process which modifies the stack, you need to activate the correct channel in the Main tab. In each channel, a stack and a mesh can be stored. They are two separate entities (saved separately) but they share the same coordinate system (move together). A stack has two stores: "Main" (never modified after being loaded) and "Work" (once it is edited, the result is overwritten). Editing stacks and meshes takes place via processes. Always make sure that the correct stack (1 or 2), store (Main or Work), and mesh are activated before running a process (displayed at the bottom right of the window). You can also modify stacks and meshes using interactive tools. To use these tools, simultaneously press Alt on your keyboard and left click with the mouse

designed to track cell growth and divisions (for instructions how to store the parent labels, *see* **Note 15**).

19. To calculate the change in cell area and display it as a heat map, you need to have T0 in Stack 1 and T1 with Parents activated in Stack 2. With Stack 2 active, run the "Mesh/Heat Map/Heat Map" process. In the dialog window, check the "Change map—Decreasing" option. You can export the values to a csv file by ticking the "Report to spreadsheet" box. You should now see a heat map on Stack 2. The values on each cell show cell area (T1)/cell area (T0). Values above 0 mean that a cell has expanded, and values below 0 mean that a cell has shrunk.

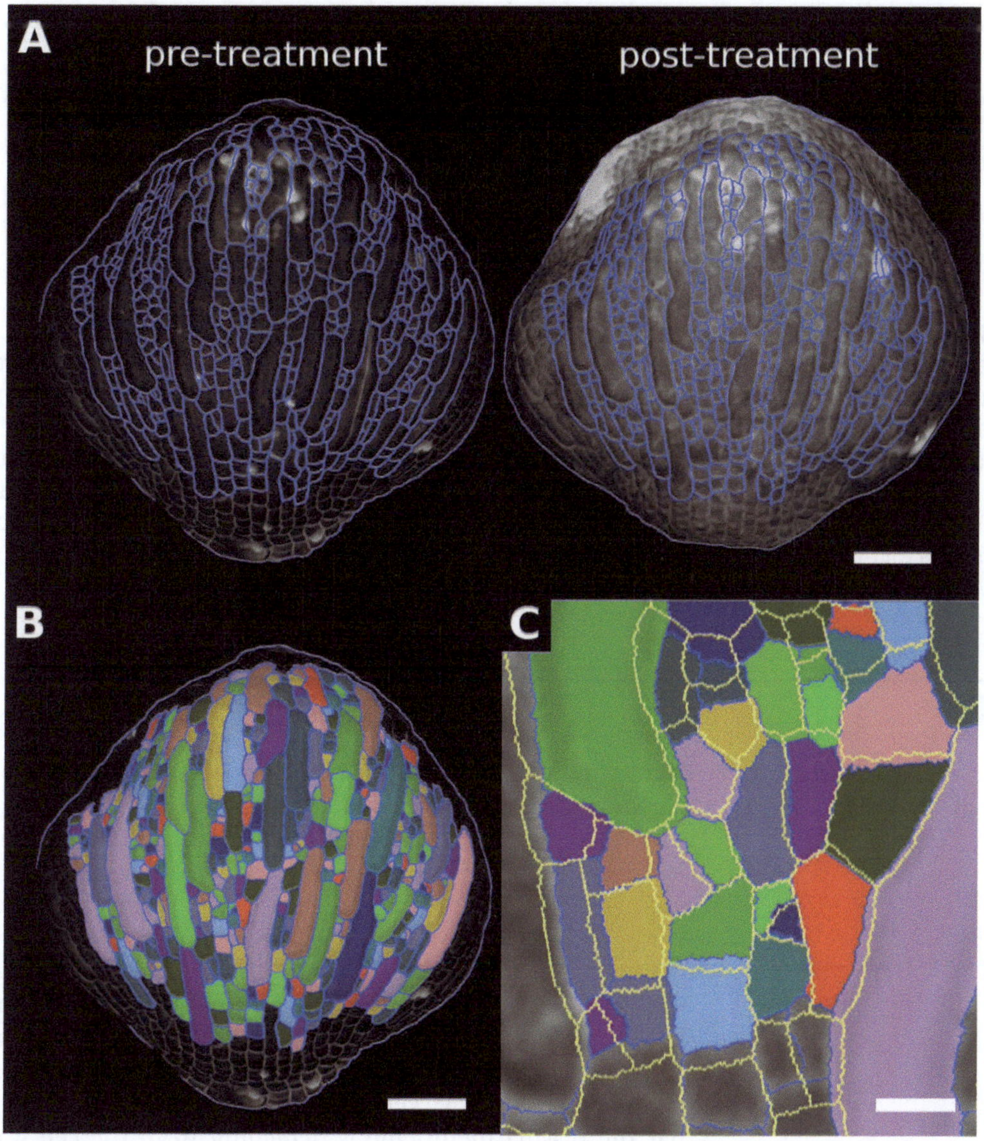

Fig. 2 Segmentation in MorphoGraphX. (**a**) Meshes built on confocal images of the sample before and after osmotic treatment, with single cell segmented (cell borders outlined in purple). (**c**) During segmentation, each cell is assigned a unique label. (**c**) In a procedure called "Parent labeling ", cell labels from the initial time point (outlined in yellow) are transferred to the cells from the final time point (outlined in purple). Scale bars: A, B, 50 μm; C, 10 μm

For instructions on how to visualize a heat map of cell shrinkage (in percentage), *see* **Note 16**.

20. Take snapshots of your results using the "Save Screenshot" tool. Compare the final results (percentage of shrinkage) in the case of segmenting single cells (Fig. 3a) and groups of cells (Fig. 3c).

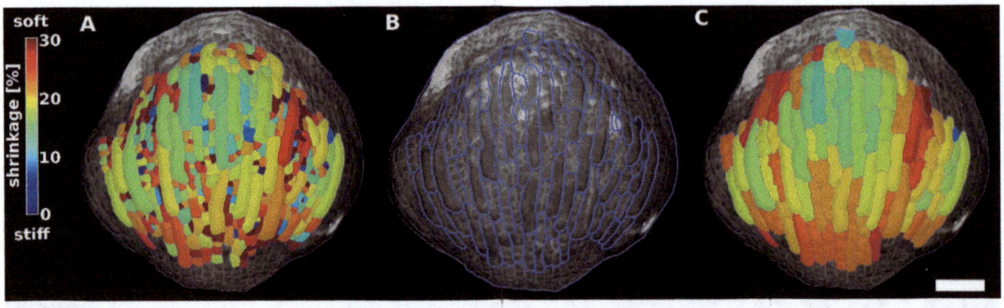

Fig. 3 Results of osmotic treatment. (**a**) Map of cell shrinkage displayed on the post-treatment time point. Note the variability in neighboring small cells. (**b**) If post-treatment image quality is not good enough, or you want to average the values for small cells, join the small cells into groups which resemble giant cells in shape and are convex. Then, run the shrinkage analysis the same way as for single cells. (**c**) Map of shrinkage for giant cells and averaged groups of small cells. Using this method, we have demonstrated that the stiffer zones of the sepal (here: the tip and center) roughly overlap with the zones that grow slower at that particular stage of flower development [6, 25]. Furthermore, we were able to detect difference in cell wall stiffness between *Arabidopsis* wild-type and *ftsh4* mutant characterized by increased reactive oxygen species levels and sepal size variability [6]. Scale bar, 50 μm

4 Notes

1. We recommend using a 0.2 M solution of NaCl for very young, meristematic cells (when imaging the shoot apical meristem). For more mature cells which have a cuticle layer, a higher concentration might be necessary. We used 0.4 M solution of NaCl as osmotic agent for sepals. However, before conducting your experiments for the first time, you should confirm if plasmolysis really occurs in your plants. Prepare a range of NaCl solutions with increasing concentrations, for example, 0.1 M, 0.2 M, 0.4 M, 0.6 M, and 1 M. Then perform the treatment as described in Subheading 3.2, and image the plasma membrane marker and PI at the same time, to confirm that the plasma membrane is detaching from the cell wall. If you would like to preserve your plants, for example, for further time-lapse work, you can use another less damaging osmotic agent such as mannitol, although it makes image acquisition more difficult.
2. If you are using more mature, differentiated, and intact tissue such as the sepal, the loss of water will be slower, and you can take the first image immediately in the osmotic agent solution, but you need to be fast. If you are using meristematic cells with little cuticle such as in the shoot apical meristem, or tissue with wounds nearby from dissected organs, it is better to take the T0 image in water after 20 min incubation and then transfer the sample into the osmotic agent solution for incubation for T1.

When in doubt, or when trying the experiment for the first time, do it in water first.

3. Imaging conditions for propidium iodide: 605–644 nm emission spectrum, 488 nm laser for excitation. Try to set the z-step as low as your microscope allows you. For MorphoGraphX analysis, image resolution in depth (z) should be the same (or possibly lower than) than in the x–y plane.
4. Make sure to rinse the microscope lens with water after this experiment. Otherwise, crystals of salt will remain on the lens.
5. Use the left click to rotate and right click to translocate the stacks. If you want to move Stack 1 or Stack 2 separately, go to the "View" tab and set "Ctrl key interaction" accordingly, then use the control key.
6. In this process, you need to adjust signal intensity threshold in order to convert the stack into a mask (solid shape full inside). This parameter will vary from sample to sample, usually between 10,000 and 20,000. The aim is to get a mask with surface as smooth as possible but, at the same time, resembling the shape of your sample as closely as possible. For images with irregular signal (or low quality) the mask might have holes, try running the "Stack/Morphology/Fill Holes" process on the mask.
7. To delete parts of the mesh, use the "Select points in mesh" tool and press delete to remove selected parts.
8. After running the subdivision, the total number of vertices will appear in the bottom left corner of the MorphoGraphX window. Depending on the power of your computer, try to limit the number of vertices less than one million. Try a few times to find the right balance between segmentation accuracy (*see* Subheading 3.3, **step 6**) and reasonable calculation times for your computer. It is also possible to adaptively subdivide to "save" vertices. Please see the MorphoGraphX user guide for more information.
9. There is no "undo button" in MorphoGraphX, so save your work frequently.
10. We recommend to fill multiple mesh triangles with the same seed within one cell, for example, by "drawing a line" in each cell. Make sure not to cross cell borders. Caution: with each mouse click, a new seed will be activated. If you want to keep putting in the same seed, change to the "Add current seed" tool or hold the Shift key as well as control.
11. For under-segmented cells: remove the label encompassing multiple cell (use the "Bucket" tool for this), put in a new seed, and rerun the "Watershed segmentation." For over-

segmented cells: pick the label of one “sub-cell” (“Pick label” tool) and fill the other sub-cell with it (“Fill label” tool).

12. In the post- treatment time point, quality of the confocal image will decrease (Fig. 2a). While putting in the seeds in the mesh, you can “draw” some cells manually using the “Add current seed” tool. However, if you cannot clearly distinguish the cell borders with your visually, it may be better to combine cells in both time points (*see* next note).
13. It might happen that borders between small cells are illegible. In this case, you will need to group several small cells into larger shapes. Try to “fit them in” between the giant cells (in the case of sepals) and keep them as convex as possible. *See* Fig. 3b. Try to keep the general pattern of cell grouping similar between compared datasets.
14. Due to the three-dimensional nature of your images, you will never be able to superimpose all cells ideally at the same time. You will need to adjust the position of Stack 1 as you go. Explore the “Scale” bars on the bottom of the “Main” tab. You can scale any stack in one or all three dimensions at the same time to make the superimposition easier.
15. The list of parent labels (from T0) corresponding to labels from T1 will be saved together with the mesh if you are using MorphoGraphX version 2.0 or later. You can also export it as a separate csv file using the “Mesh/Lineage Tracking/Save Parents” process, and then load it back using the “Mesh/ Lineage Tracking/Load Parents” process. Once you have your parent labels loaded on T1, you can switch the view from T1 labels to T0 parent labels by ticking and unticking the “Parents” box in the Mesh section.
16. If you want to see the percentage value of how much your cells shrunk, you need to modify the csv file containing the heat map externally. We recommend to do it in R, MatLab, or MS Excel. Replace the values X in the “Value” column with $(1 - X) * 100$, and save the csv file under a new name. Then, use the “Mesh/Heat Map/Heat Map Load” process to load the shrinkage values. The final result can be seen in Fig. 3a.

Acknowledgment

We thank Daniel Kierzkowski for guidance in tissue dissection and Gabriella Mosca and Mingyuan Zhou for comments on the manuscript.

References

1. Forouzesh E, Goel A, Mackenzie SA, Turner JA (2013) In vivo extraction of *Arabidopsis* cell turgor pressure using nanoindentation in conjunction with finite element modeling. Plant J 73:509–520
2. Beauzamy L, Derr J, Boudaoud A (2015) Quantifying hydrostatic pressure in plant cells by using indentation with an Atomic Force Microscope. Biophys J 108:2448–2456
3. Milani P, Gholamirad M, Traas J, Arnéodo A, Boudaoud A, Argoul F et al (2011) In vivo analysis of local wall stiffness at the shoot apical meristem in *Arabidopsis* using Atomic Force Microscopy. Plant J 67:1116–1123
4. Kierzkowski D, Nakayama N, Routier-Kierzkowska A-L, Weber A, Bayer E, Schorderet M et al (2012) Elastic domains regulate growth and organogenesis in the plant shoot apical meristem. Science 335:1096–1109
5. Cosgrove DJ (2005) Growth of the plant cell wall. Nat Rev Mol Cell Biol 6:850–861
6. Hong L, Dumond M, Tsugawa S, Sapala A, Routier-Kierzkowska A-L, Zhou Y et al (2016) Variable cell growth yields reproducible organ development through spatiotemporal averaging. Dev Cell 38:15–32
7. Elsayad K, Werner S, Gallemí M, Kong J, Guajardo ERS, Zhang L et al (2016) Mapping the subcellular mechanical properties of live cells in tissues with fluorescence emission—Brillouin imaging. Sci Signal 9:1–13
8. Peaucelle A, Braybrook SA, LeGuillou L, Bron E, Kuhlemeier C, Höfte H (2011) Pectin-induced changes in cell wall mechanics underlie organ initiation in *Arabidopsis*. Curr Biol 21:1720–1726
9. Sampathkumar A, Krupinski P, Wightman R, Milani P, Berquand A, Boudaoud A et al (2014) Subcellular and supracellular mechanical stress prescribes cytoskeleton behavior in *Arabidopsis* cotyledon pavement cells. elife 3: e01967. https://doi.org/10.7554/eLife.01967
10. Beauzamy L, Louveaux M, Hamant O, Boudaoud A (2015) Mechanically, the shoot apical meristem of *Arabidopsis* behaves like a shell inflated by a pressure of about 1 MPa. Front Plant Sci 6:1–10
11. Majda M, Grones P, Sintorn IM, Vain T, Milani P, Krupinski P, Zagorska-Marek B, Viotti C, Jonsson H, Mellerowicz E, Hamant O, Robert S (2017) Mechanochemical polarization of contiguous cell walls shapes plant pavement cells. Dev Cell 43:290–304
12. Routier-Kierzkowska A-L, Weber A, Kochova P, Felekis D, Nelson BJ, Kuhlemeier C et al (2012) Cellular force microscopy for in vivo measurements of plant tissue mechanics. Plant Physiol 158:1514–1522
13. Hayot CM, Forouzesh E, Goel A, Avramova Z, Turner J (2012) Viscoelastic properties of cell walls of single living plant cells determined by dynamic nanoindentation. J Exp Bot 63:2525–2540
14. Bolduc J-E, Lewis LJ, Aubin C-E, Geitmann A (2006) Finite-element analysis of geometrical factors in micro-indentation of pollen tubes. Biomech Model Mechanobiol 5:227–236
15. Wang L, Hukin D, Pritchard J, Thomas C (2006) Comparison of plant cell turgor pressure measurement by pressure probe and micromanipulation. Biotechnol Lett 28:1147–1150
16. Park YB, Cosgrove DJ (2012) A revised architecture of primary cell walls based on biomechanical changes induced by substrate-specific endoglucanases. Plant Physiol 158:1933–1943
17. Mosaliganti KR, Noche RR, Xiong F, Swinburne I, Megason SG (2012) ACME: Automated Cell Morphology Extractor for comprehensive reconstruction of cell membranes. PLoS Comput Biol 8(12):e1002780. https://doi.org/10.1371/journal.pcbi.1002780
18. Robinson S, Huflejt M, Barbier de Reuille P, Braybrook S, Schorderet M, Reinhardt D et al (2017) An automated confocal micro-extensometer enables in vivo quantification of mechanical properties with cellular resolution. Plant Cell 29:2959–2973
19. Bringmann M, Bergmann DC (2017) Tissue-wide mechanical forces influence the polarity of stomatal stem cells in *Arabidopsis*. Curr Biol 27:877–883
20. Weber A, Braybrook S, Huflejt M, Mosca G, Routier-Kierzkowska AL, Smith RS (2015) Measuring the mechanical properties of plant cells by combining micro-indentation with osmotic treatments. J Exp Bot 66:3229–3241
21. Barbier de Reuille P, Routier-Kierzkowska A-L, Kierzkowski D, Bassel GW, Schüpbach T, Tauriello G et al (2015) MorphoGraphX: a platform for quantifying morphogenesis in 4D. elife 4:05864. https://doi.org/10.7554/eLife.05864
22. Oparka KJ (1994) Plasmolysis: new insights into an old process. New Phytol 67:571–591

23. Mosca G, Sapala A, Strauss S, Routier-Kierzkowska AL, Smith RS (2017) On the micro-indentation of plant cells in a tissue context. Phys Biol 14:015003. https://doi.org/10.1088/1478-3975/aa5698
24. Roeder AHK, Chickarmane V, Cunha A, Obara B, Manjunath BS, Meyerowitz EM (2010) Variability in the control of cell division underlies sepal epidermal patterning in Arabidopsis thaliana. PLoS Biol 8:e1000367. https://doi.org/10.1371/journal.pbio.1000367
25. Hervieux N, Dumond M, Sapala A, Routier-Kierzkowska AL, Kierzkowski D, Roeder AHK et al (2016) A mechanical feedback restricts sepal growth and shape in *Arabidopsis*. Curr Biol 26:1019–1028
26. Smyth DR, Bowman JL, Meyerowitz EM (1990) Early flower development in *Arabidopsis*. Plant Cell 2:755–767

Chapter 12

Mapping a Transcriptome-Guided *Arabidopsis* SAM Interactome

Muhammad Naseem, Ozge Osmanoglu, Jibran Iqbal, Fares M. Howari, Fatima A. AlRemeithi, Martin Kaltdorf, and Thomas Dandekar

Abstract

The advent of multi-OMICS approaches has a significant impact on the investigation of biological processes occurring in plants. RNA-SEQ, cellular proteomics, and metabolomics have added a considerable ease in studying the dynamics of stem cell niches. New cell sorting approaches coupled with the labeling of stem cell population specific marker genes are highly instrumental in enriching distinct cellular populations for various types of analysis. One more promising field of OMICS is the mapping of cellular interactomes. The plant stem cells research is barely profited from this newly emerging field of OMICS. Generation of stem cell/niche-specific interactome is a time-consuming and labor-intensive task. Here, we describe a method on how to assemble a SAM-based interactome after using the available generic *Arabidopsis* interactomes. To define the context of SAM in a generic interactome, we used SAM cell population transcriptome datasets. Our step-by-step protocol can easily be adopted for other stem cell niches such as RAM and lateral meristems keeping in view the availability of transcriptome datasets for cellular populations of these niches.

Key words SAM, RAM, Transcriptome, OMICs, Interactomes

1 Introduction

By default, plants are programmed to replenish lost body organs such as flowers, leaves, and external protective layers on regular basis. To constantly supply precursor cells that grow into either of the mentioned organs requires a source of cells to regenerate the lost body parts [1]. The pluripotent stem cells constantly supply precursor cells to form differentiated tissues and body organs [2]. There are three stem cell niches in plants: the shoot apical meristem (SAM), the root apical meristem (RAM), and the vascular meristem. The stem cell niches maintain a spatial signaling block to avoid them from entering differentiation zones all at once and keep a required amount of undifferentiated stem cells [2, 3]. The SAM is

Muhammad Naseem and Ozge Osmanoglu contributed equally to this work.

Muhammad Naseem and Thomas Dandekar (eds.), *Plant Stem Cells: Methods and Protocols*, Methods in Molecular Biology, vol. 2094, https://doi.org/10.1007/978-1-0716-0183-9_12, © Springer Science+Business Media, LLC, part of Springer Nature 2020

a true custodian of the next generation of plants because it constantly supplies cells to meet the programmed and contingency requirements of the aerial parts of the plants [2, 3].

The SAM is a dynamic structure of a hemispherical collection of similarly appearing cells. It provides a stable organization that maintains a balance between the self-renewal of stem cell population and conversion of meristematic cells that are destined to become differentiated cells and committed tissues [4, 5]. For the sake of convenience, the SAM *Arabidopsis* can be divided into three regions: The central zone (CZ) is at the tip of the SAM and harbors pluripotent stem cells. A collection of multipotent stem cells derived from the CZ constitutes the peripheral zone (PZ), from which the primordia of leaves and flowers come are formed [1, 2]. The rib meristem lies beneath the central and PZs; it turns into cells of the stem and vasculature. Plant hormone cytokinins and auxin are believed to be the key signaling mediators in maintaining the integrity of the SAM stem cell niche [1, 3]. The patchy distribution of these two hormones plays a significant role in SAM-signaling networks.

The SAM cell populations are divided into numerous forms with various potencies and occur in different zones; however, the domains of various proteins that define the boundaries of various subcellular populations are always fused, and hence it is very changeling to demarcate a specific cell population in a specific zone such as CZ and PZ. The prominent cell populations that can be isolated on the basis of specific cellular markers are WUS-expressing cells, AtHB8 layer of cells, FIL, S17, CLV3p-expressing cells (stem cells), HGD layer, LAS cell layer, KAN1, and HMG layers of the SAM cellular populations. With the advent of OMICS in plant stem cells research, now it is possible to isolate single and subject them to whole-transcriptome analysis. Similarly, proteome and metabolic dissection of the various SAM cell populations is more in routine these days. To better analyze SAM-based signaling and developmental processes, it is direly important to have an idea about the SAM protein-protein interaction (interactomes) networks. Generally cellular interactomes are very much generic and cannot be extrapolated from organ or tissue or cell type to another cell type. Here we combined a generic *Arabidopsis* interactome with SAM-specific gene expression (transcriptome, *see* Fig. 1) date and came up with a SAM-based transcriptome-guided interactome for model plant *Arabidopsis*. This will better serve the plant stem cell community in order to investigate various developmental and signaling processes occurring in the SAM.

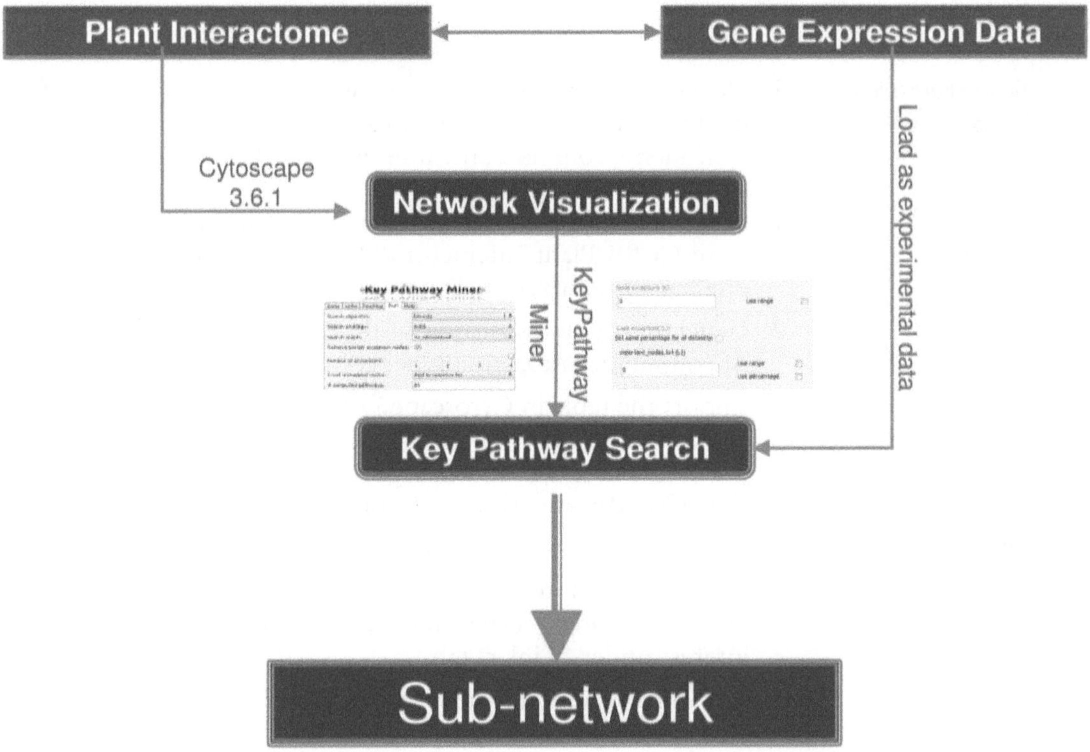

Fig. 1 Overview of the methods used. Interactome data used to build a main network is investigated for key pathways connecting important nodes that are determined from the gene expression data

2 Materials

1. CCSB Plant Interactome Database (http://interactome.dfci.harvard.edu/A_thaliana/index.php?page=download): *Arabidopsis* interactome database.
2. Gene expression profiles of SAM populations from Gene Expression Omnibus [6, 7].
3. Geo-2R, a statistical package that normalizes cellular transcriptome in a user-friendly manner [8].
4. Cytoscape3.6.1 (https://cytoscape.org/): Network Data Integration, Analysis, and Visualization software [9].
5. KeyPathwayMiner (https://keypathwayminer.compbio.sdu.dk/keypathwayminer/): condition-specific pathway analysis tool [10].

3 Methods

3.1 Gene Expression Data Preparation

1. Generate a table consisting of two columns: column 1, gene name (should be in the same format with node names or gene identifiers); column 2, the number 1 (*see* Fig. 1 **Notes 1** and **2**).

3.2 Network Visualization

1. Download the *Arabidopsis* protein-protein interaction dataset from CCSB Plant Interactome Database (http://interactome.dfci.harvard.edu/A_thaliana/index.php?page=download).
2. Generate a table consisting of three columns: column 1, node A; column 2, interaction type; column 3, node B.
3. Import the table in Cytoscape3.6.1 (File > Import > Network > File).

3.3 KeyPathwayMiner

1. Open KeyPathwayMiner App on Cytoscape3.6.1.
2. Load the expression data under Data tab.
3. If there are more than one experimental dataset: upload all separately under "Data" tab; specify the logical connector of interest under "Links" tab.
4. Add other important nodes of interest based on prior knowledge under "Pos/Neg" tab.
5. Under "Run" tab, choose the parameters of interest, search algorithm, and search strategy. Specify a value for K(Node exceptions) which will be taken into account when searching for a key pathway as the maximum number of non-important nodes the pathway can have. Set a value for L if more than two datasets are loaded.
6. Search key pathways (*see* Fig. 1).

4 Notes

1. Analyzed microarray data [6] is first filtered using a threshold of p-value < 0.05 and $|\log2FC| > 1$ and then combined to obtain a list of DEGs enriched in 10 samples spanning different cell layers and populations residing in central and peripheral zones of SAM (*see* Fig. 2a).
2. With this approach, all genes that are differentially regulated in any of the populations or cell layers are included in the final list. The list is converted to a table with two columns that is later fed as experimental dataset into the Cytoscape plug-in KeyPathwayMiner. An example table is shown below (*see* Table 1).

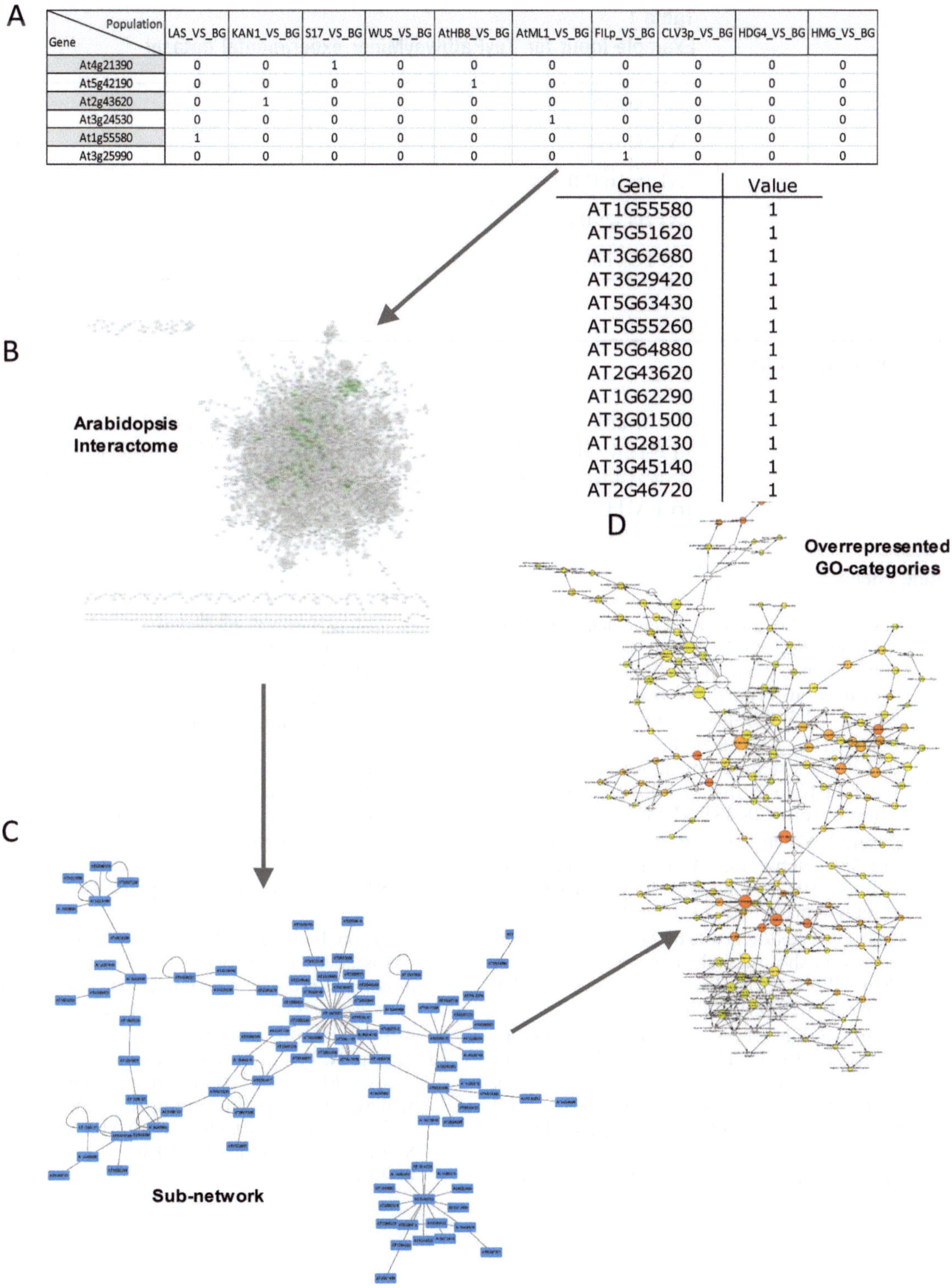

Fig. 2 Example output from the key pathway search. (**a**) Gene expression data assembled into a table showing upregulated (white) or downregulated (gray) genes in different SAM cell populations, further put into a table regardless of their direction of regulation to be employed in key pathway search. (**b**) The key pathway (green) found in *Arabidopsis* interactome (gray) after analysis with KeyPathwayMiner. (**c**) The key pathway viewed separately, consisting of 95 nodes and 108 edges

Table 1
Example input for KeyPathwayMiner experimental dataset

AT3G28600	1
AT2G39850	1
AT3G16470	1
AT2G37900	1
AT1G11580	1

Acknowledgment

We thank the German Research Foundation (DFG) for funding (TR124/B1) to TD and start-up grant (R18045) and RIF-grant by Zayed University to MN and UAE Space Agency grant (EU1804) to FMH.

References

1. Naseem M, Srivastava M, Dandekar T (2014) Stem-cell-triggered immunity safeguards cytokinin enriched plant shoot apexes from pathogen infection. Front Plant Sci 5:588. https://doi.org/10.3389/fpls.2014.00588
2. Aichinger E, Kornet N, Friedrich T, Laux T (2012) Plant stem cell niches. Annu Rev Plant Biol 63:615–636. https://doi.org/10.1146/annurev-arplant-042811-105555
3. Hwang I, Sheen J, Müller B (2012) Cytokinin signaling networks. Ann Rev Plant Biol 63:353–380. https://doi.org/10.1146/annurev-arplant-042811-105503
4. Perales M, Reddy GV (2012) Stem cell maintenance in shoot apical meristems. Curr Opin Plant Biol 15:10–16. https://doi.org/10.1016/j.pbi.2011.10.008
5. Song X-F, Yu D-L, Xu T-T et al (2012) Contributions of individual amino acid residues to the endogenous CLV3 function in shoot apical meristem maintenance in arabidopsis. Mol Plant 5:515–523. https://doi.org/10.1093/mp/ssr120
6. Yadav RK, Tavakkoli M, Xie M et al (2014) A high-resolution gene expression map of the Arabidopsis shoot meristem stem cell niche. Development 141:2735–2744. https://doi.org/10.1242/dev.106104
7. Edgar R, Domrachev M, Lash AE (2002) Gene expression omnibus: NCBI gene expression and hybridization array data repository. Nucleic Acids Res 30:207–210. https://doi.org/10.1093/nar/30.1.207
8. Barrett T, Wilhite SE, Ledoux P et al (2013) NCBI GEO: archive for functional genomics data sets—update. Nucleic Acids Res 41: D991–D995. https://doi.org/10.1093/nar/gks1193
9. Shannon P, Markiel A, Ozier O et al (2003) Cytoscape: a software environment for integrated models of biomolecular interaction networks. Genome Res 13:2498–2504. https://doi.org/10.1101/gr.1239303
10. Alcaraz N, Pauling J, Batra R et al (2014) KeyPathwayMiner 4.0: condition-specific pathway analysis by combining multiple omics studies and networks with Cytoscape. BMC Syst Biol 8:99. https://doi.org/10.1186/s12918-014-0099-x

Chapter 13

3D Analysis of Mitosis Distribution Pattern in the Plant Root Tip with iRoCS Toolbox

Viktoriya V. Lavrekha, Taras Pasternak, Klaus Palme, and Victoria V. Mironova

Abstract

The protocol allows to define and characterize mitosis distribution patterns in the plant root meristem. The method does not require genetic markers, which makes it applicable to plants of different non-transgenic genotypes, including ecotypes, mutants, and non-model plant species. Computer analysis of the mitosis distribution in three dimensions with iRoCS Toolbox identifies statistically significant changes in proliferation activity within specific root tissues and cell lineages.

Key words RAM, Proliferation, Cell cycle, Image analysis, *Arabidopsis*, Epidermis, Cortex, Endodermis, Pericycle, Protophloem, Protoxylem

1 Introduction

Plant scientists usually consider the root meristem as an organ of simple anatomy, wherein concentric layers of the epidermis, cortex, endodermis, and pericycle encircle the vascular cylinder. However, root tissues (e.g., the protophloem, protoxylem, and metaxylem) start specializing inside the root meristem [1]. 3-Dimensional analysis of mitosis distribution in the root tip allows defining the earliest steps in the meristematic tissue specialization. Analysis of mitosis distribution in *Arabidopsis* root meristem showed that cells of different lineages exit the cell cycle at different distances from the Quiescent Center (QC) with the protophloem lineage exiting first and procambium last [2].

Root meristem length is an important trait showing root growth potential. The conventional criterion to estimate the meristem length in *Arabidopsis* is the first elongating cell of the cortex [3, 4]. However, at least in *Arabidopsis* this marker underestimates

Viktoriya V. Lavrekha and Taras Pasternak contributed equally to this work.

Muhammad Naseem and Thomas Dandekar (eds.), *Plant Stem Cells: Methods and Protocols*, Methods in Molecular Biology, vol. 2094, https://doi.org/10.1007/978-1-0716-0183-9_13, © Springer Science+Business Media, LLC, part of Springer Nature 2020

the size of the meristem, because cortex stops dividing earlier than procambium [2]. Another way to estimate proliferation activity in the root meristem is to use CYCB1 (CYCLIN B1) as genetic marker. CYCB1;1-GUS or CYCB1;1-GFP chimeric genes are widely used to visualize the cells in G2/M transition [5]. However, CYCB1;1 expresses in a tissue-specific manner [2]; therefore, using this marker does not always guarantee correct results. Recently, 5-ethynyl-20-deoxy-uridine (EdU) has been proposed to monitor DNA replication in vivo [6]. At least in tobacco root tips, it was shown that the domain of EdU-positive cells covers both the root meristem and the distal elongation zone [7], implying that EdU also should be used with caution when analyzing proliferation activity in tissue.

Here we propose an unbiased method to study proliferation activity in plant root tips by direct observation of mitosis distribution. The experimental-computational method applies to study three-dimensional mitosis distribution over different plant tissues and cell lineages.

2 Materials and Methods

This protocol was developed for *Arabidopsis thaliana* plants (Fig. 1); however, it was successively applied to study mitosis distribution in tobacco [7], tomato (Fig. 2), and wheat (*see* **Note 1**). All computer programs used for image analysis are open source.

2.1 Plant Materials and Growth Conditions

1. Surface-sterilize and sow seeds on square Petri dishes containing AM medium (½ strength Murashige and Skoog medium containing vitamins, 1% sucrose, and 1% w/v agar).
2. Keep the plates at room temperature for 4 h before transfer to 4 °C for 12 h.
3. Transfer the plates to 22 °C under a 16 h:8 h light:dark photoperiod with a light intensity of 80 μmol·m^{-2}·s^{-1} for 4.5 days.
4. Transfer the seedlings to a 6-well plate containing liquid T1 medium [8].
5. After a 16 h preincubation in T1 medium, add 2.5 mM colchicine and incubate the seedlings for a further 90 min (*see* **Note 2**).

2.2 Plant Materials and Growth Conditions

1. Fix plants in 4% formaldehyde in microtubule stabilizing buffer (MTSB).
2. Wash twice with distilled water for 10 min.
3. Incubate in 200 μg·l^{-1} DAPI for 20 min.
4. Wash again with distilled water.

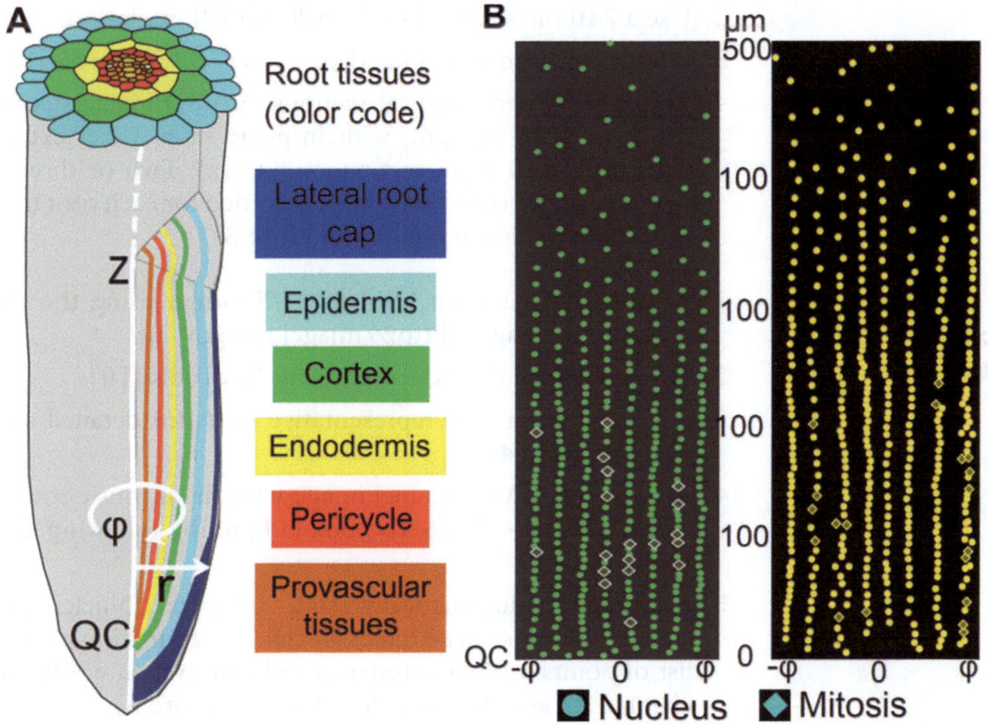

Fig. 1 iRoCS Toolbox implementation for mitosis annotation in the root tip in three dimensions. iRoCS Toolbox provides the position of each nucleus in the cylindrical coordinate system (**a**) that allows analyzing root structure layer by layer. Implementation of cylindrical coordinates gives the advantage to unroll tissue layers (**b**): cortex and endodermis

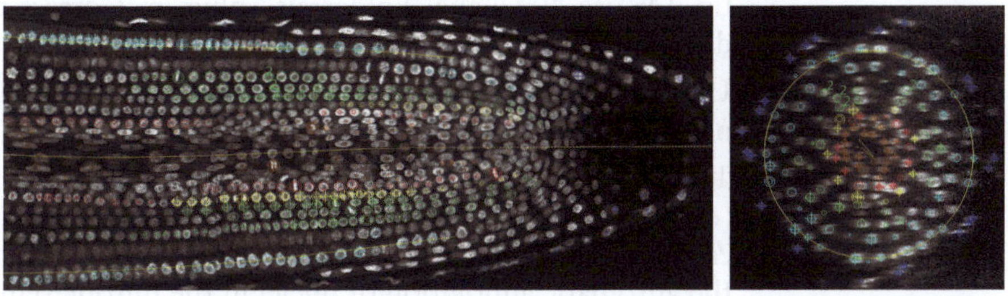

Fig. 2 Annotation of tomato root tip using iRoCS Toolbox [10]. Plants were grown and analyzed according to current protocol; the root tips were mounted with 240 μm spacer. Nuclei and mitoses were annotated manually

5. Mount on slides with a 120 μm spacer for *Arabidopsis* and 240–360 μm for other species using ProLong® Gold Antifade Mountant (ThermoFisher Scientific, Waltham, MA, USA) [9].

2.3 Confocal Imaging

1. Scan DAPI-stained samples using a confocal microscope, for example, LSM 510 META NLO; Carl Zeiss, Oberkochen, Germany, with a LD LCI Plan-Apochromat 25×/0.8 DIC Imm Corr objective.

2. Use a 740 nm Chameleon laser for DAPI excitation.
3. Detect emission with a band-pass filter (BP 390–465 IR).
4. Reconstitute serial optical sections into 3D image stacks to a depth of 120–360 μm, with in-plane (*x*–*y*) voxel extents of 0.15 μm and 0.9 μm section spacing (*z*). Two or three overlapping images (tiles) should be recorded for each root to cover the whole meristem region (*see* **Note 3**).

2.4 Image Processing in iRoCS Toolbox

1. Convert images from LSM to *hdf5* format using the "hdf5" plug-in for ImageJ (http://imagej.nih.gov/ij).
2. Stitch confocal tiles together using XuvTools [10].
3. Choose at least four representative roots for detailed annotation (*see* **Note 4**).
4. Process the DAPI channel image with the intrinsic root coordinate system—iRoCS Toolbox [11] in the following way (*see* **Note 5**).
5. Detect nuclei automatically using "01-Detect Nuclei" plug-in (menu "Plugins→01-Detect Nuclei." The function generates a list of points in the Cartesian coordinate system written in the channel "/ annotation / detector" (*see* **Note 6**).
6. Manually detect the location of the QC nuclei using the function "Channel→New Annotation Channel." Create a new channel "QC," which contains only one point—the QC location.
7. Manually detect 20–60 epidermal nuclei along the entire root tip. Or use "02-Label Epidermis" plug-in. Correct the autodetection if necessary.
8. Use "03-Attach iRoCS" plug-in to set the cylindrical coordinates to all nuclei of the root.
9. To get automatic classification of the nuclei to the cell types (epidermis, endodermis, cortex, pericycle, vasculature, root cap), use the "04-Assign Layers" plug-in, which also enables the automatic annotation of nuclei in the mitotic state (option "Re-classify mitotic state") (*see* **Note 7**).
10. Correct manually all erroneous layer and mitosis assignments if necessary (*see* **Note 8**).
11. One can mark additional features using "Marker Control" frame, "Subtype," or "Cell file" fields. For example, to specify protophloem nucleus, edit the marker subtype from "Unspecified" to "3 (phloem)" (*see* **Note 9**). Any other features, e.g., "protophloem pole" or "periclinal division," can be annotated by editing "Cell file" field.

2.5 Data Analysis

1. Export the data from iRoCS Toolbox to .csv files ("Channel→Export annotation channel to CSV..."). Exported data contains the coordinate of each nucleus in the Cartesian and cylindrical coordinate systems. This data allows to "unroll" concentric root layers and analyze their nuclei positioning and labels more precisely (Fig. 2).
2. Use the data from "label," "subtype," and "cellfile" columns to select the coordinates of nuclei belonging to specific tissues or lineages.
3. Use the data from "distance from QC (z) (micron)" column to determine the distances from the QC to the mitotic nuclei in the tissue or lineage. Sort the data to get the coordinates of the final mitosis.
4. Also you can define mitosis distribution (MD) along the central axis in an individual tissue or in the whole root as the proportion of mitoses inside 50 μm interval relative to the number of mitoses over the studied tissue.
5. To analyze statistical differences, we recommend using Welch's t-test or nonparametrical Mann-Whitney u-test.

3 Notes

1. The only limitation of this method is the thickness of the root tip.
2. Short incubation period was chosen to avoid the adverse effects of the treatment (detailed in [11]). At this step, the method can be complimented with EdU labeling to visualize the nuclei in both mitotic and replication/endoreduplication steps.
3. Final mitoses occur in the provascular tissues far away from the first elongating cortex cell, so make more tiles in advance. In case of additional EdU labeling, we recommend to scan the root tip including the distal elongation zone. In *Arabidopsis* 4 dag plants, we scanned the root tips of 500 μm in length.
4. As more roots were scanned and annotated, as better statistics you get. Optimal number is 10–15 roots.
5. If there are other channels, then load them separately. During annotation, you can use several channels at the same time.
6. More information about iRoCS Toolbox is here: https://lmb.informatik.uni-freiburg.de/lmbsoft/iRoCS/#usage. See there the system requirements, examples of usage, and functionality of the Toolbox.
7. Colchicine treatment (Subheading 2.1, **item 5**) leads to anaphase failure, so the mitotic nuclei can be detected as compact objects with dense DAPI signal.

8. Depending on the quality of the confocal image and the root tip anatomy, one should do more or less manual corrections. In case of poor quality images or in case of abnormal *Arabidopsis* root tip anatomy, we recommend to do the annotation manually. "04-Assign Layers" plug-in uses the model developed for *Arabidopsis*; if one annotates the root tips of other plant species, using this plug-in does not guarantee correct results. However, it is possible to create a new model for automatic layer assignment in the root tips of your plant species (*see* iRoCS Toolbox tutorial).
9. It is not easy to annotate the nuclei of the vascular cylinder: provascular tissues are clearly visible when wall-stained roots, but not when nuclei-stained. However, one can distinguish protophloem by the nuclei phenotype: when specializing protophloem loses the nucleus [12]. When protophloem lineages are mapped, one can determine protoxylem and metaxylem cells on a central perpendicular plane to the protophloem poles.

Acknowledgment

We thank Roland Nitschke and the staff of the Life Imaging Center (LIC) in the Center for Biological Systems Analysis (ZBSA) of the Albert-Ludwigs-University Freiburg for help with their confocal microscopy resources. This work was supported by the Russian State Budget Project (0324-2019-0040), by the Russian Federation President Grant for young scientists (MK-1297.2017.4), and by the Russian Foundation for Basic Research (RFBR-18-34-00485). This work was also supported by Bundesministerium für Bildung und Forschung (BMBF Microsystems, Haploswitch), the German Research Foundation (DFG SFB746, INST 39/839,840,841), and Deutsches Zentrum für Luft und Raumfahrt (DLR 50WB1022). The authors declare that no competing interests exist.

References

1. Bishopp A, Lehesranta S, Vaten A, Help H, El-Showk S et al (2011) Phloem-transported cytokinin regulates polar auxin transport and maintains vascular pattern in the root meristem. Curr Biol 21:927–932
2. Lavrekha VV, Pasternak T, Ivanov VB, Palme K, Mironova VV (2017) 3D analysis of mitosis distribution defines the longitudinal zonation and bilateral symmetry of the Arabidopsis thaliana root meristem. Plant J 92 (5):834–845
3. Casamitjana-Martınez E, Hofhuis HF, Xu J, Liu CM, Heidstra R, Scheres B (2003) Root-specific CLE19 overexpression and the sol1/2 suppressors implicate a CLV-like pathway in the control of Arabidopsis root meristem maintenance. Curr Biol 13(16):1435–1441
4. Blein T, Duerr J, Pasternak T, Haser T, Falk T et al (2018) Light dynamically regulates growth rate and cellular organisation of the Arabidopsis root meristem. bioRxiv 1:353987
5. Colon-Carmona A, You R, Haimovitch-Gal T, Doerner P (1999) Technical advance: spatio-

temporal analysis of mitotic activity with a labile cyclin-GUS fusion protein. Plant J 20 (4):503–508

6. Hayashi K, Hasegawa J, Matsunaga S (2013) The boundary of the meristematic and elongation zones in roots: endoreduplication precedes rapid cell expansion. Sci Rep 3:2723
7. Pasternak T, Haser T, Falk T, Ronneberger O, Palme K, Otten L (2017) A 3D digital atlas of the Nicotiana tabacum root tip and its use to investigate changes in the root apical meristem induced by the Agrobacterium 6b oncogene. Plant J 92(1):31–42
8. Pasternak TP, Rudas VA, Lörz H, Kumlehn J (1999) Embryogenic callus formation and plant regeneration from leaf base segments of barley (Hordeum vulgare L.). J Plant Physiol 155:371–375
9. Pasternak T, Tietz O, Rapp K, Begheldo M, Nitschke R, Ruperti B et al (2015) Protocol: an improved and universal procedure for whole-mount immunolocalization in plants. Plant Methods 11:1
10. Emmenlauer M, Ronneberger O, Ponti A, Schwarb P, Griffa A, Filippi A et al (2009) XuvTools: free, fast and reliable stitching of large 3D datasets. J Microsc 233:42–60
11. Schmidt T, Pasternak T, Liu K, Blein T, Aubry-Hivet D et al (2014) The iRoCS Toolbox—3D analysis of the plant root apical meristem at cellular resolution. Plant J 77:806–814
12. Truernit E, Bauby H, Belcram K, Barthélémy J, Palauqui J-C (2012) OCTOPUS, a polarly localised membrane-associated protein, regulates phloem differentiation entry in Arabidopsis thaliana. Development 139:1306–1315

Chapter 14

Micropropagation of Rosaceous Species SAM Grown in Temperate Climate

Jiri Sedlak and Frantisek Paprstein

Abstract

The benefits of in vitro plant cultivation are mainly due to very high multiplication rate. Cultivation of plant material in vitro can be carried out during the whole year regardless of the time of the year or weather conditions. We create artificial conditions in the lab (heat, light, humidity), and we can regulate these conditions at any time. For the preservation of cultivar identity, we recommend establishing in vitro cultures from shoot tips usually larger than 0.2 mm. In practice, in vitro cultivation of plants uses these growth regulators to achieve organogenesis, for example, root formation, prolonged growth, or multiplication. During each subculture, these cultures are then transferred on a solid agar medium in the form of actively growing multiple shoots with a well-differentiated shoot tip containing meristematic area. Cytokinins are important for cell division and causes branching of plants. Auxins, both endogenous and exogenous, act at as a trigger for the differentiation and formation of root primordia. Morphological characteristics (formation of leaves or callus) and shoot development should be observed during in vitro multiplication and after transfer to ex vitro conditions.

Key words In vitro, Multiplication rate, Growth regulator, Media, Explant

1 Introduction

Today tissue cultures of fruit species are exploited for various purposes. The benefits of in vitro plant cultivation are mainly due to very high multiplication rate. If suitable genotypes are to receive wide distribution, rapid propagation methods are very important to provide supplies of large numbers of high-quality plants to end users. The generative propagation of most fruit species in a commercial scale is hardly usable, because it does not give stable and morphologically homogeneous progeny. Most fruit plants can be propagated by vegetative methods (tillers, cuttings, budding). Although propagation by vegetative means is generally successful, it is slow and labor intensive. Other limiting factors are dependence upon the season and insufficient rooting and growth vigor in the case of some genotypes. Plant material produced in vitro also

Muhammad Naseem and Thomas Dandekar (eds.), *Plant Stem Cells: Methods and Protocols*, Methods in Molecular Biology, vol. 2094, https://doi.org/10.1007/978-1-0716-0183-9_14, © Springer Science+Business Media, LLC, part of Springer Nature 2020

exhibits growth uniformity as it is multiplied by the vegetative means (meri-clone production) [1–3]. Saving of space is also a significant advantage. Cultivation of plant material in vitro can be carried out during the whole year regardless of the time of the year or weather conditions. We create artificial conditions in the lab (heat, light, humidity), and we can regulate these conditions at any time. Another important use is the possibility of elimination of plant pathogens, especially viral diseases by chemotherapy, in vitro thermotherapy, or using the so-called Cryo knife during freezing in liquid nitrogen.

In vitro cultivation is also widely used for scientific purposes, for example, in experimental biology (e.g., determination of heavy metal effects on plant species, phytohormone concentrations and various plant-based substances, GMOs, etc.). Plant explants can be also used for industrial production of specific substances (secondary metabolites—substances of pharmaceutical importance, fragrances, essential oils, pigments, substances with antimicrobial activity etc.) [4–6]. In vitro cultivation of plants is also an issue associated with plant breeding (embryo cultures of early sweet cherries).

The inconsistent results of previous research on in vitro cultivation continue to motivate further micropropagation studies. The underlying physiological and biochemical mechanisms [7] that determine the need for growth regulators during the different stages of in vitro development are currently being investigated in many scientific workplaces [8]. There is no universal medium for rosaceous fruit species in vitro culture, since cultivars and genotypes are genetically specific with regard to different components of the medium.

2 Material

1. A suitably equipped chemical laboratory containing at least analytical scales as well as precision scales.
2. A pH meter, a laboratory stirrer, a refrigerator, and a microwave must be available.
3. For sterilization of solutions and agar growing media, a sterilization device at a temperature of at least 120 °C for that steam pressure autoclave is required (*see* Fig. 1).
4. For the sterilization of work tools such as scalpels and tweezers, a hot air dryer is suitable, which uses dry hot air at specified temperature and time parameters (*see* Fig. 1).
5. In addition, a device for the preparation of demineralized water by reverse osmosis or electro-distillation is recommended. This water is used for the preparation of cultivated, agar-solidified soils (*see* Fig. 1).

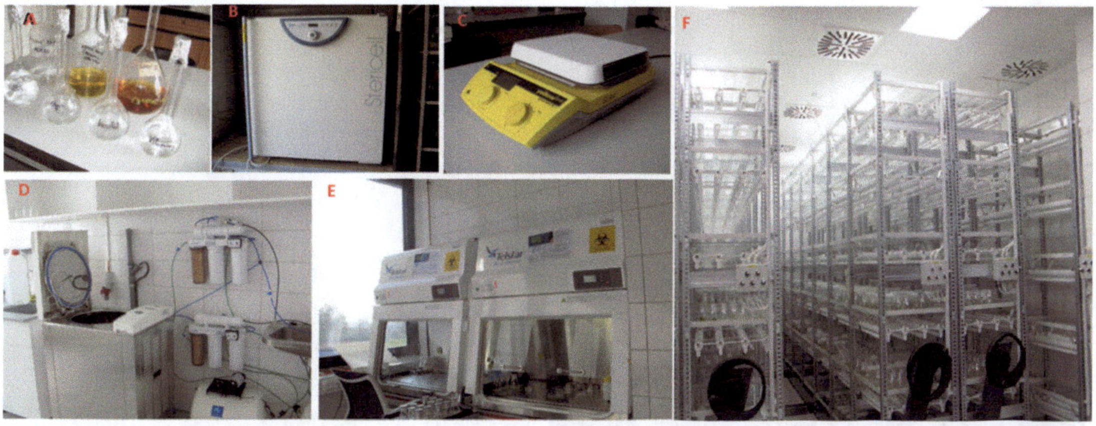

Fig. 1 Essential equipment and infrastructure required for the development of plant apex-derived disease-free varieties. (**a**) Stock solutions and media types. (**b**) Hot air dryer and magnetic stirrer. (**c**) Device for preparation of demineralized water by reverse osmosis and steam pressure autoclave. (**d**) Aseptic flow box—biohazard type. (**e**) Aseptic flow box—laminar flow type. (**f**) High-capacity cultivation room for the multiplication of the developed explants

6. For the growth of in vitro cultures, a temperature- and lighting-controlled culture room is required (*see* Fig. 1).
7. Passage and subsequent transfer of in vitro plants to fresh media should be performed in an aseptic flow box with flowing filtered air. The flow box allows the material to be kept sterile while handling (sub-cultivation, *see* Fig. 2) outside the culture vessel.
8. Volumetric cylinders, volumetric flasks, Erlenmeyer flasks, beakers, pipettes, sprays, stirrers, syringes, antibacterial microfilters, etc. are required to prepare stock solutions and media (*see* Fig. 1).
9. Scalpels, tweezers, glass Petri dishes, scissors, and aluminum foils are required to work with sterile plant material.
10. As culture vessels, we use mainly Erlenmeyer flasks (100–250 mL), glasses sealable with a sterilizable transparent lid or specially developed plastic containers for in vitro cultivation.
11. Composition of modified MS medium according to Murashige a Skoog (*see* Table 1).
12. Cytokinins should be used in a geometric sequence (e.g., 1, 2, and 4 mg l^{-1} BAP or 0.5, 1, and 2 mg l^{-1} TDZ (thidiazuron)) in order to process shoot formation statistically.

Fig. 2 Establishment of in vitro culture, multiplication, and transfer to ex vitro conditions. (**a**) Established in vitro explants of plum. (**b**)Multiplying well-developed pear in vitro shoots from MS medium—BAP (2 mg l^{-1}). (**c**) Multiplying pear explant from MS medium with TDZ (1 mg l^{-1})—excessive callus formation at the base of explant. (**d**) Example of acclimatization to ex vitro conditions E. (**e**) Rooted acclimatized plants in "Jiffy" pots

Table 1
Composition of modified MS medium according to Murashige a Skoog [6]

Compound	mg · l^{-1}	Compound	mg · l^{-1}
NH_4NO_3	1650	$CuSO_4 \cdot 5H_2O$	0.025
KNO_3	1900	$Na_2EDTA \cdot 2H_2O$	37.3
H_3BO_3	6.2	$FeSO_4 \cdot 7H_2O$	27.8
KH_2PO_4	170	Thiamin	0.1
KI	0.83	Pyridoxine	0.5
$Na_2MoO_4 \cdot 2H_2O$	0.25	Nicotinic acid	0.5
$CoCl_2 \cdot 6H_2O$	0.025	Glycin	2
$CaCl_2 \cdot 2H_2O$	440	Sucrose	30,000
$MgSO_4 \cdot 7H_2O$	370	Myo-inositol	100
$MnSO_4 \cdot 4H_2O$	22.3	Agar	7000
$ZnSO_4 \cdot 7H_2O$	8.6	pH	5.8

13. Auxins indole-3-butyric acid (IBA), indole-3-acetic acid (IAA), and naphthaleneacetic acid (NAA) at concentration 0.5–2 mg l^{-1} are the most common plant growth regulators to trigger root induction.
14. Demonstration of proliferation rates—pear cultivars Butterbirne Lukas, Milada, and Red Bartlett.

3 Method

3.1 In Vitro Culture Establishment of Initial Plant Material

For the preservation of cultivar identity, we recommend establishing in vitro cultures from shoot tips usually larger than 0.2 mm. During each subculture, these cultures are then transferred on a solid agar medium in the form of actively growing multiple shoots with a well-differentiated shoot tip containing meristematic area. For practical in vitro multiplication, we do not recommend using less organized callus or cell culture less than 0.1 mm due to risk of genetic instability. It facilitates sterilization and establishment of uncontaminated initial explants. The success of sterilization procedure is proportional to the lowest possible level of contamination. In addition to the dormant period, actively growing shoot tips can be taken and prepared directly from field conditions, preferably at the beginning of vegetation (from mid-April to mid-June) during periods of intensive shoot elongation.

3.2 Sterilization

When establishing in vitro culture, we recommend washing of shoots under pure running water. This will remove a part of microbial flora that survives on the surface of shoots in outdoor conditions and subsequently can cause contamination after the in vitro culture is established. The apical parts of shoots are prepared for sterilization procedure by removing all expanded leaves, leaf parts, and other surface structures preventing sterilization. It can be carried out also under a stereoscopic microscope. The tips are immersed in 0.5% sodium hypochlorite or 0.15% mercuric chloride (be careful highly toxic!). Several drops of wetting agent (Tween 20) can be added to sterilization solutions. Wetting agent is important because it facilitates the penetration of the sterilizing agent into the potentially contaminated area. The sterilization time is 1 min, followed by three rinses with sterile distilled water. The sterilized explants are placed to cultivation vessels with culture medium. The culture media for rosaceous species include usually MS salts and vitamins [6] supplemented with 100 mg L^{-1} inositol, 2 mg L^{-1} glycin, 30 g L^{-1} sucrose, and 1.5 mg L^{-1} 6-benzylaminopurine (BAP). The medium is gelled with 0.7% (W/v) high-quality agar. The pH of the medium is adjusted to 5.8 before adding the agar and autoclaving at 120 °C at 100 kPa for 15 min. Established cultures are incubated in a growth room under 16 h of cool-white fluorescent light provided by cool-white tubular lamps at

Table 2
Example—results of surface sterilization of pear by 0.15% $HgCl_2$

	Contaminated explants		Uncontaminated explants which did not develop shoots		Uncontaminated explants which developed shoots	
Genotype	Number	(%)	Number	(%)	Number	(%)
Butterb. Luk.	1	5.0	0	0.0	19	95.0
Milada	0	0.0	1	5.0	19	95.0
Red Bart.	0	0.0	6	30.0	14	70.0

22 ± 1 °C. The typical irradiance should be 40–60 μmol m^{-2} s^{-1} at plant height. In order to detect possible contamination by undesirable microorganisms, it is necessary to visually monitor in vitro cultures for at least 1 month. Developing shoots are serially subcultured onto fresh media for 4–10 consecutive 4-week passages. This provides a stock collection of shoots for proliferation studies (Table 2).

3.3 Proliferation Experiments

We recommend testing different proliferation MS media containing various concentrations of cytokinins in a geometric sequence (e.g., 1, 2, and 4 mg l^{-1} BAP or 0.5, 1, and 2 mg l^{-1} TDZ (thidiazuron)). Thermolabile cytokinins like TDZ should be filter sterilized (0.2 μm membrane) and added to proliferation media after autoclaving. Uniform developing shoots (10–15 mm in length including the apex) are detached from previously cultured explants and transferred to fresh medium for shoot proliferation. After 30 days, the number of newly formed shoots, callus formation, and shoot morphology should be determined for each genotype and concentration of particular growth regulator. The response of individual genotypes can significantly vary during the micropropagation. Proliferation rate can be defined as the number of newly formed shoots (>10 mm) per initial shoot tip after four weeks of culture. The shoot formation should be recorded between the seventh and tenth subculture. Each combination of genotype and treatment should involve at least 10 shoot tips, and each experiment should be repeated at least three times. Data from independent experiments are pooled and expressed as the mean. We recommend to compare treatment means with the standard error (SE) of the mean. The standard error of the mean describes a dispersion of sample means around the population mean. Morphological characteristics (formation of leaves or callus) and shoot development should also be observed (Table 3).

3.4 Root Induction Using Auxins In Vitro Culture Conditions

Once enough shoots are available from studied genotypes, their rooting ability should be tested. Auxins, as growth regulators, play an important role in management of plant cell division, elongation,

Table 3
Example of different shoot proliferation rates for pear genotypes ± SE

Cytokinin (mg l^{-1})	Butterbirne Lukas	Milada	Red Bartlett
BAP			
1	1.9 ± 0.1	2.2 ± 0.1	2.2 ± 0.1[a]
2	4.1 ± 0.2	2.8 ± 0.2	2.4 ± 0.1[a]
4	5.1 ± 0.3	3.9 ± 0.2	2.0 ± 0.1[a]
TDZ			
0.5	3.5 ± 0.2[b]	3.2 ± 0.1	3.3 ± 0.2[a,b]
1	3.3 ± 0.2[b]	2.1 ± 0.1	2.3 ± 0.1[a,b]
2iP			
10	1.3 ± 0.1	1.7 ± 0.1	1.1 ± 0.0

[a]Undeveloped narrow leaves
[b]Short shoots with excessive callus formation at the base of explants

and differentiation. Auxins, both endogenous and exogenous, act at a sufficient concentration as a trigger for the differentiation and formation of root primordia during initial stages of root development. These root primordia differentiate from meristematic areas (growth centers) in in vitro culture conditions approximately 6 days after exposure of plant explant to auxin. There are significant differences in the in vitro rooting ability of individual fruit crop genotypes. Single cultured shoots (15–20 mm in length) are excised from the best proliferation media and placed upright in MS medium with lower concentration of macro- and micronutrients (at around 50%). Auxins indole-3-butyric acid (IBA), indole-3-acetic acid (IAA), and naphthaleneacetic acid (NAA) at concentration 0.5–2 mg l^{-1} are the most common plant growth regulators to trigger root induction. We recommend the addition of auxins to the finished medium via antibacterial micro-filters after sterilization. Shoots should be exposed to in vitro rooting treatment for 2–6 weeks. Culture conditions during root initiation and root growth are the same as during shoot culture. Rooting can be expressed as percentage of shoots rooting and number of roots per rooted culture 1 month after transfer to rooting medium. We recommend to compare treatment means for number of roots with the standard error (SE) of the mean (Table 4).

3.5 Acclimatization to Ex Vitro Conditions

Plants produced in rooting experiment must be rinsed in water to remove remnants of the agar medium and misted with water to prevent wilting. In vitro rooted shoots should be transferred to peat "Jiffy" pots (e.g., Jiffy 7, AS Jiffy Products, Norway) soaked with water. After transfer, the "Jiffy" pots with rooted plants are placed

Table 4
Example of different rooting potential of pear cultivars on 50% MS medium in descending order

Auxin cultivar	Rooting shoots (%)	Root number per shoot ± SE
NAA (1 mg l^{-1})		
Butterbirne Lukas	63	2.0 ± 0.2
Red Bartlett	59	1.9 ± 0.1
Milada	26	2.7 ± 0.9
IBA (1 mg l^{-1})		
Red Bartlett	59	3.3 ± 0.3
Butterbirne Lukas	42	1.8 ± 0.2
Milada	4	1.0 ± 0.0
IAA (1 mg l^{-1})		
Red Bartlett	5	1.0 ± 0.0
Butterbirne Lukas	4	1.0 ± 0.0
Milada	0	0.0 ± 0.0

in trays under clear plastic humidity domes and incubated under light in growth chambers at 22 °C. For better acclimatization, covers were opened gradually during another 2-week period. Acclimatized plants should be potted and transferred to the greenhouse, where they need to be observed for morphological characteristics.

4 Notes

1. Macroelements and microelements form an inorganic component of artificial media that is essential for the growth and development of plant material. During media preparation, they are added in the form of inorganic salts. The plants receive these elements in the form of ions from the medium. Their uptake by plants from the medium is influenced by concentration of other elements, pH, by the temperature in cultivation room, etc. Macroelements include nitrogen, phosphorus, potassium, calcium, magnesium, and sulfur. Microelements used for in vitro plant cultivation are iron (physiologically macroelement), zinc, boron, copper, manganese, cobalt, and molybdenum. The separate application is then in the form of prepared stock solutions of mixtures of microelements or macroelements or other compounds. Many of these elements can be included in mixtures for preparation of nutrient media produced by commercial companies, which makes the laboratory preparation itself easier and faster.

2. Plant material cultured in artificial in vitro conditions needs energy to develop, grow, and multiply. Sucrose is the most common source of this energy. Sucrose inhibits formation of chlorophyll and photosynthesis, and plants change to a heterotrophic diet. In addition, it also adjusts the osmotic value of the medium. Most often, the sucrose concentration is 2.5–4.0% (i.e., 30 g sucrose per liter of medium). In some cases, sucrose may be replaced by other sugars such as sorbitol, fructose, glucose, raffinose, maltose, lactose, mannitol, etc.
3. Plant hormone (phytohormone or plant growth regulator) is an organic compound that is produced by the plant itself in one part and then transferred to another, where it produces a certain physiological response. Both natural and synthetic growth regulators can stimulate or inhibit the plant in certain concentration. In practice in artificial laboratory conditions, in vitro cultivation of plants uses these growth regulators to achieve organogenesis, for example, root formation, prolonged growth, or multiplication of plants. The basic and most frequently used phytohormones in plant cultivation under in vitro conditions are cytokinins, auxins, gibberellins, abscisic acid, and ethylene.
4. Auxins are synthesized in apex, young leaves, organs, developing fruits, and seeds. Its content is high mainly in young, intensely growing organs and decreases with aging of particular plant organ. Auxin levels are very low in xylem of vascular plants. Auxins are essential for organogenesis and plant formation and maintaining of plant polarity. Auxins are responsible for apical dominance. Together with cytokinins, they affect branching of aboveground and underground plant parts. In in vitro laboratory practice, they are used for rooting induction and elongation of stem. Auxins indole-3-butyric acid (IBA), indole-3-acetic acid (IAA), and naphthaleneacetic acid (NAA) at concentration 0.5–2 mg l^{-1} are the most common plant growth regulators to trigger root induction.
5. Cytokinins are important for hormonal regulation of cell division of explants and cell cultures. They have the opposite effect as auxins—they reduce apical dominance and cause branching of plants. Together with auxins, they allow plants to regenerate in explant cultures. Today, over 30 natural cytokinins are known, of which kinetin (6-furfulaminopurine) and zeatin are the best known. The synthetic cytokinins used in rosaceous fruit species micropropagation include, for example, 6-benzylaminopurine (BAP) or thidiazuron (TDZ). In conclusion, cytokinins are mainly used as a component of culture media for the multiplication of plant material.

6. In vitro culture media can be prepared as solid or liquid. The solid consistency of the medium is most often due to the addition of agar. Agar is a nonhomogeneous mixture of polysaccharides that is prepared from various red seaweed (e.g., genus *Acanthopeltis, Ceramium, Gelidium, Gracilaria, Pterocladia*). For conventional plant cultivation, 5–7 g of agar per liter of medium is used. Agar also contains a small amount of macro- and microelements (e.g., K, P, Ca, Fe, Mn, Na, Mg, Zn), sugars, vitamins, and amino acids. Agar is the most commonly used medium for solidifying other substances. Agars of varying degrees of purity are available at market.

References

1. Bell RL, Reed BM (2002) In vitro tissue culture of pear: advances in techniques for micropropagation and germplasm preservation. Acta Hortic 596:412–418
2. Durkovic J (2006) Rapid micropropagation of mature wild cherry. Biol Plant 50(4):733–736
3. Erbenova M, Paprstein F, Sedlak J (2001) In vitro propagation of dwarfed rootstocks for sweet cherry. Acta Hortic 560:477–480
4. Hossini AD, Moghadam EG, Anahid S (2010) Effects of media cultures and plant growth regulators in micro propagation of gisela 6 rootstock. Ann Biol Res 1(2):135–141
5. Mansseri-Lamrioui A, Louerguioui A, Bonaly J, Yakoub-Bougdal S, Allili N, Gana-Kebbouche S (2011) Proliferation and rooting of wild cherry: the influence of cytokinin and auxin types and their concentration. Afr J Biotechnol 10 (43):8613–8624
6. Murashige T, Skoog F (1962) A revised medium for rapid growth and bioassays with tobacco tissue cultures. Physiol Plant 15:473–497
7. Reed BM, DeNoma J (2016) Medium-term in vitro storage of pear as a complementary germplasm preservation technique. Acta Hortic 1113:251–256
8. Ruzic D, Vujovic TI (2008) The effects of cytokinin types and their concentration on in vitro multiplication of sweet cherry cv. Lapins (*Prunus avium* L.). HortSci 35(1):12–21

Chapter 15

A New Perspective on Cryotherapy: Pathogen Elimination Using Plant Shoot Apical Meristem via Cryogenic Techniques

Ergun Kaya, Selin Galatali, Sevinc Guldag, and Onur Celik

Abstract

Plant pathogens cause different diseases on crops and industrial plant species that result in economic losses. Pathogen-free plant material has usually been obtained by traditional procedures such as meristem culture, thermotherapy, and chemotherapy. However, there are many limitations of these procedures such as mechanical challenges of meristem excision and low regeneration rate, low resistance to high temperatures, phytotoxicity, and mutagenic effects of the chemicals used in the procedures. Cryotherapy is a newly developed biotechnological tool that has been very effective in virus elimination from economically important plant species. This tool has overcome the abovementioned limitations. This chapter aims to highlight the importance of the cryogenic procedures (vitrification, encapsulation-vitrification, droplet vitrification, two-step freezing, dehydration, encapsulation-dehydration) in order to generate virus-free germplasm.

Key words Plant biotechnology, Dehydration, Liquid nitrogen, Plant viruses, Vitrification

1 Introduction

Plant pathogens such as bacteria, fungus, and viruses cause harmful diseases on plants, and some of them can cause direct and/or indirect losses of billions of dollars every year. Plant viruses causing diseases on plants can destroy crops and industrial plant species; therefore, they have negative effect on food security and crop industry [1, 2]. Chemical therapies or physical treatments are not sufficient to be directly controlled of them. There are many different traditional ways to prevent for virus contaminations such as biological and chemical control of the vector being often an insect transmitting viruses, growing virus resistant crop varieties being made via genetic transformations, using virus-free planting material and the protection of disease placement in fields where viruses do not yet occur [3, 4].

Muhammad Naseem and Thomas Dandekar (eds.), *Plant Stem Cells: Methods and Protocols*, Methods in Molecular Biology, vol. 2094, https://doi.org/10.1007/978-1-0716-0183-9_15, © Springer Science+Business Media, LLC, part of Springer Nature 2020

Pathogen-free (especially virus-free) plant material has generally been obtained using meristem culture and/or thermotherapy methods [5, 6]. For successful viral elimination via meristem culture, it is usually necessary to excise shoot tips that are 0.1–0.5 mm (depend on plant species) in size [7, 8]. There are some limitations for virus elimination using these methods such as mechanical challenges of meristem excision and low regeneration rate [9]. Thermotherapy method connected with meristem culture is also a difficult process requiring specific conditions such as virus-specific treatments and specific equipment. However, all of viruses cannot be eliminated via thermotherapy, and sometimes the infected plants are not resistant to high temperatures [10].

An alternative method is chemotherapy for plant virus elimination. This method based on the usage of antiviral chemicals associated with thermotherapy or meristem culture was successfully used for virus elimination of some infected plants such as apple [10]. The antiviral chemicals such as quercetin and ribavirin prevent virus nucleic acid synthesis (replication), and thus the virus concentration cannot increase in infected plants [11]. However, phytotoxicity and mutagenic effects of these antiviral chemicals are reported for some plant species and/or cultivar [12].

Because of some limitations of the traditional methods, it is beneficial to develop different kind of efficient biotechnological procedures for obtaining virus-free plants. Cryotherapy—newly developed biotechnological tool—has been very effective for virus elimination for economical important plant species such as sweet potato [13], strawberry [14], raspberry [15], potato [16], grape [17], and apple [18].

2 Cryotherapy

Pathogen-free (virus-free) plant materials are the most important for agricultural and horticultural crop productivity and for ornamental plant quality [19]. The plants especially vegetatively propagated are inclined to pathogen infections, which are transported to new plants in infected steels, tubers, roots, and other vegetative parts of plants. Conservation of plant genetic resources is one of the most important tools for breeding new species and plant cultivars for future requirements. However, germplasm collections need to be established from pathogen-free species and cultivars. Therefore, development of efficient methods for pathogen elimination is a critical point of gene banks for maintenance of their collections [20].

Cryotherapy is a new method used for pathogen elimination from infected plant shoot tips [13]. There are many reports for successfully pathogen elimination from plants infected by different kind of virus and bacteria like pathogens via cryotherapy of shoot

Table 1
Different cryotherapy treatments for pathogen elimination for vegetatively propagated and economically important plant species [21]

Plant	Pathogen	Cryotherapy method	Reference
Banana (Musa)	*Cucumber mosaic virus* (CMV)/ *Banana streak virus* (BSV)	Vitrification	[22]
Beijing lemon, mandarin, pummelo, sweet orange (*Citrus*)	*Huanglongbing bacterium* (HLB)	Vitrification	[23]
Grapevine (*Vitis vinifera*)	*Grapevine virus* (GVA)	Encapsulation-vitrification	[24]
Grapevine (*Vitis vinifera*)	*Grapevine virus* (GVA)	Encapsulation-dehydration	[25]
Potato (*Solanum tuberosum*)	*Potato leaf roll virus* (PLRV)/*Potato virus* Y (PVY)	Encapsulation-vitrification Droplet vitrification	[26]
Prunus hybrid	*Plum pox potyvirus* (PPV)	Vitrification	[27]
Raspberry (*Rubus idaeus*)	*Raspberry bushy dwarf virus* (RBDV)	Thermotherapy followed by cryotherapy (Encapsulation-vitrification)	[8]
Sweet potato (*Ipomoea batatas*)	*Sweet potato chlorotic stunt virus* (SPCSV)/*Sweet potato feathery mottle virus* (SPFMV)	Encapsulation-vitrification	[13]
Yam (*Dioscorea opposita*)	*Yam mosaic virus* (YMV)	Encapsulation-dehydration	[28]

tips, such as potato [16] and sweet potato [13]. Cryotherapy can be used for a wide range of plant species and cultivars because it is based on plant cryopreservation methods being available for many additional vegetatively propagated and economically important plant species (Table 1).

In cryotherapy technique, infected plant cells are eliminated by the fatal efficacy of liquid nitrogen (−196 °C, the ultralow temperature) and/or following warming; mechanical removal is not required. After cryotherapy treatments, shoot tip regeneration rates might be lower than traditional meristem culture treatments; however, larger shoot tips can be used for easier excision, and obtained pathogen-free materials are much more via cryotherapy [13].

2.1 Cryogenic Procedures

Cryotherapy involving physical dehydration and chemical vitrification treatments of shoot tips are not needed special equipment in addition to that used in a basic plant tissue culture laboratory.

Liquid nitrogen being the main material for cryotherapy is usually easy available in laboratories found in almost all countries. Cryogenic treatments are divided into two major groups including traditional techniques (classical slow cooling) and one-step freezing techniques [29].

2.1.1 Traditional Techniques: Classical Slow Cooling

Classical slow cooling methods cover two-step cooling down to a specified prefreezing temperature (−40 °C), followed by direct immersion in liquid nitrogen. During reduction of temperature by slow cooling, cells and the other medium firstly supercool, followed by ice formation in the medium [30]. The plant cell membrane behaves as a physical barrier and prevents the fatal ice nucleation from the cell inner, and the cells stay unfrozen but supercooled. As the temperature is further reduced, an increasing volume of the extracellular solution is transformed into ice form, in this way resulting in the concentration of intracellular solutes. whereby the plant cells continue supercooled and hydrated vapor pressure of them passes over that of the frosted other compartment, plant cells balance via loss of water to other ice formation. Before the intracellular components solidify, different amounts of cell water content will be removed depending upon the prefreezing temperature and the cooling rate. In ideal conditions, big volume or almost all intracellular water causing ice nucleation is removed, in this way decreasing or preventing fatal intracellular ice nucleation during liquid nitrogen immersion. But sometimes dehydration causing more intense ice nucleation can induce a series of damaging cases due to intracellular salt concentration and modifies in the plant cell membrane [31]. Rewarming process should be as fast as possible to prevent the fatal re-ice nucleation cases in which ice reforms at a thermodynamically suitable, bigger, and more harm ice nucleation form [30].

Classical slow cooling processes contain different consecutive steps: cold hardening (pre-cold culture at +4 °C) and sucrose preculture (on preculture medium supplemented with different concentrations of sucrose) of plant materials, cryoprotection (chemical vitrification or physical dehydration), slow cooling (0.5–1 °C/min) to transferring a prefreezing temperature (approximately −80 °C), immersion of samples rapidly in liquid nitrogen, storage process, rapid rewarming, and recovery. Classical slow cooling methods are usually operationally complicated since they require the use of specific and costly programmable freezers. Sometimes, it can be used cheap and a specific tool named Mr. Frosty® freezing container based on usage of propanol (allows 1 °C/min temperature reduction) with a -80 laboratory freezer [32, 33].

Classical slow cooling techniques have been successfully applied to many plant culture types especially in cell suspension cultures and callus cultures [33, 34].

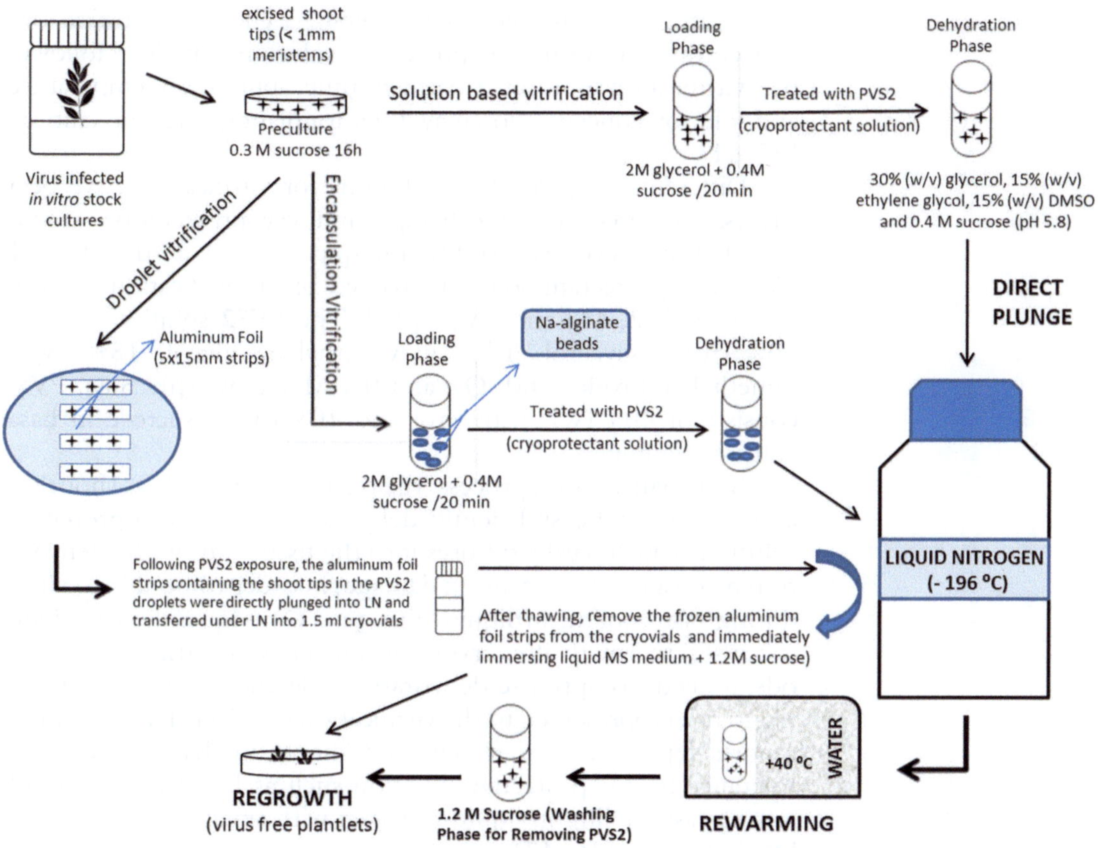

Fig. 1 Schematic presentation of three different vitrification methods based on one-step freezing techniques: cryotherapy, vitrification, encapsulation-vitrification, and droplet vitrification [21]

2.1.2 One-Step Freezing Techniques

One-step freezing techniques can be achieved via direct immersion in liquid nitrogen without slow cooling process (Fig. 1), which is changed by exposure of plant materials to physical dehydration or chemical vitrification using a cryoprotectant solution. This technique is divided into five main procedures: vitrification, dehydration, encapsulation vitrification, encapsulation-dehydration, and droplet vitrification [29].

Vitrification

Vitrification processes are based on the physical treatments, of which a high concentration of cryoprotectant solution prevents fatal ice nucleation in cells during direct immersion in liquid nitrogen [35]. Because of amorphous glass formation of cell water content, all metabolic reactions requiring molecular diffusion stop, and this amorphous formation leads to metabolic inactivity and stability during immersion of liquid nitrogen [36].

Vitrification method based on cryoprotectant solutions combines a classical cooling procedure (combining the cryoprotectant treatment and dehydration steps). After cold hardening and sucrose preculture, in cryoprotectant treatment, a chemical solution of a concentrated diffusing cryoprotectant solution is applied, followed by a vitrification solution. Treatment time, application temperature, and solution concentration may differ for plant species and cultivars [37–41].

Different cryoprotectant solutions for vitrification have been successfully used for cryotherapy and cryopreservation studies [42, 43]. However, two of them frequently used are the glycerol-based cryoprotectant solutions named plant vitrification solution 2 (PVS2) [42, 44] and PVS3 [45]. The PVS2 solution contains 30% (w/v) glycerol, 15% (w/v) ethylene glycol, 15% (w/v) dimethyl sulfoxide (DMSO), and 0.4 M sucrose (pH 5.8). PVS3 consists of 40% (w/v) glycerol and 40% (w/v) sucrose in basal culture medium.

In the vitrification process, the plant material such as shoot tips and cells must be sufficiently dehydrated by the cryoprotectant solution (which hardly diffuses into the tissue during the dehydration process) without causing damage, in order to be able to vitrify during fast cooling in liquid nitrogen. Consequently, to achieve successful regrowth after cryopreservation using vitrification methods, it needs to optimize dehydration tolerance of the plant material to be cryopreserved to the vitrification solution. There are many reports representing that cells and shoot tips have dehydration tolerance to cryoprotectant solution such as PVS2 resistance following fast cooling in liquid nitrogen with small or no additional loss in survival [46, 47].

Dehydration

Dehydration method based on physical process using activated silica or laminar airflow for removing water content of tissues and cells is a very simple procedure, and it only consists of explant dehydration, then freezing them in fast direct immersion in liquid nitrogen. Zygotic embryos or embryonic parts extracted from seeds are usually used for explant sources in this technique. And this technique has also been applied to a large number recalcitrant and intermediate species [48–50]. Dehydration process is usually performed in a sterile laminar airflow cabinet; however, more specific and effective dehydration conditions are achieved by using a flow of sterile compressed air or silica gel. Ultrafast drying by using a compressed dry airstream allows freezing of samples with a relatively high water content, thus decreasing desiccation damages [51]. The water content of tissue and cells reducing between 10 and 20% (basis on fresh weight) supports optimal survival, and regrowth rate is generally obtained when samples are frozen during liquid nitrogen treatment [50].

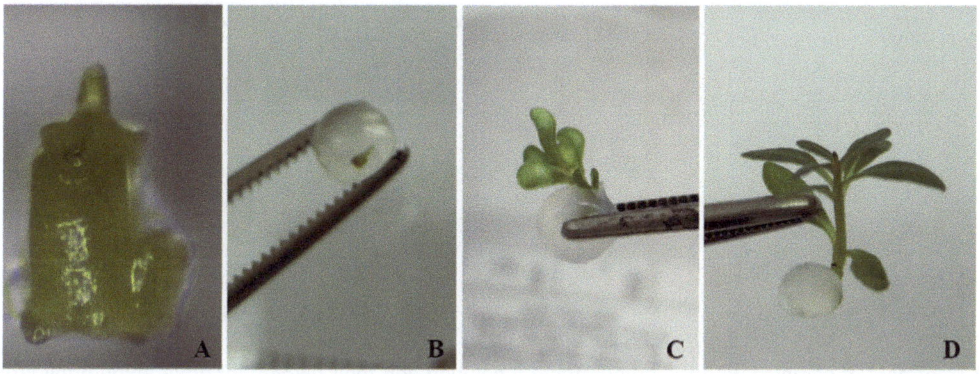

Fig. 2 Naked 0.1 mm shoot tip (**a**) and encapsulated meristem (**b**) of *Eucalyptus* spp. for cryotherapy via encapsulation-vitrification method; regrowth stages of after 4 weeks (**c**) and 6 weeks (**d**) of encapsulated meristems of frozen *Eucalyptus* spp. [38]

Encapsulation-Vitrification

Vitrification process allows the explants freezing in a short period of time. However, this process is difficult to apply for a large number of samples at the same time, as the duration of the consecutive stages of a vitrification protocol is usually too short; on the one hand, these stages require a very definite period, and small-sized plant materials are difficult to manipulate. On the other hand, the encapsulation-dehydration methods take a much longer time to perform; however, encapsulated plant materials are very easy to manipulate, by using optimum size of the calcium alginate beads (Fig. 2a–d). Thus, encapsulation-vitrification method combines the advantages of vitrification having fast application and encapsulation-dehydration having easy manipulation of encapsulated plant materials [52].

Encapsulation-Dehydration

The encapsulation-dehydration method is based on physical dehydration process of encapsulated plant materials. This method includes similar application with simple dehydration process; however, the main difference from simple dehydration is usage of encapsulated shoot tips. Explants encapsulated in calcium alginate beads desiccates in a laminar airflow cabinet or with activated silica gel for reducing water content, and then they are fast immersed directly in liquid nitrogen [53]. This technique has been used for shoot tips of many species from tropical and subtropical origin as well as to cell suspensions and somatic embryos of several species [54].

Droplet Vitrification

The droplet vitrification method based on chemical vitrification and one-step freezing treatments was first reported by Schäfer-Menuhr et al. [55] using potato shoot tips. In this technique, 01–03 mm explants (Fig. 3a) are treated with the cryoprotectant solution (usually PVS2) put individually in 3–10 μL droplets of cryoprotectant solutions (depending on explant size) placed on a piece of

Fig. 3 Treatment of samples with the cryoprotectant solution and their placement on aluminum foil strip with subsequent transfer on regrowth medium

aluminum foil strip (Fig. 3b), which is then directly immersed in liquid nitrogen. For rewarming, the aluminum foils are directly plunged in liquid medium supplemented with 1–1.2 M sucrose, and after 20 min of unloading, shoot tips are transferred on regrowth medium (Fig. 3c). The main advantage of this method is the possibility of achieving very high cooling/warming rates due to the very small volume of cryoprotectant solution which the explants are placed. Although this is a newly developed technique, there are many reports obtained with a high regrowth percentage after immersion of liquid nitrogen [38, 39, 56–58].

3 Confirmation of Pathogen-Free Plants

3.1 Immunodiagnostic Techniques: ELISA

Determination of plant viruses is usually based on their biological characteristics (host range, typical symptoms), and this process has been achieved via serological tests since the 1960s. Serological laboratory tests were originally developed for determination of viruses by using antibodies to detect epitopes of protein antigens. The immunological diagnostic methods include enzyme-linked immunosorbent assay (ELISA), immunofluorescence (IF), and immuno-strip tests [59]. ELISA is by far the most traditional used immunodiagnostic method for virus determination since the 1970s [60]. Variations on this technique exist that differ from each other in the way the antigen-antibody complex is detected, but the underlying mechanism is the same.

3.2 DNA-Based Techniques: Reverse Transcriptase-PCR (RT-PCR)

Plant virus diagnostics and detection of polymerase chain reaction-based techniques have progressively been used in recent years to improve diagnostic assays for plant pathogens. These techniques have the potential to be very sensitive and highly specific and are based on the unique nucleic acid sequence of the pathogens [61]. Cheap and effective nucleic acid extraction methods have already been described, including total RNA (Fig. 4), double-

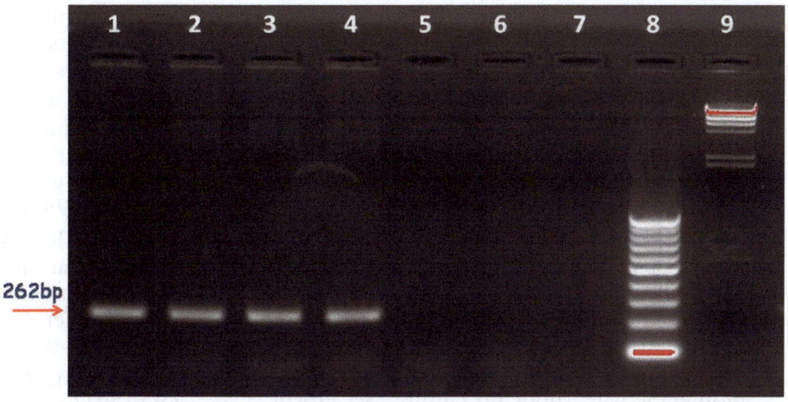

Fig. 4 Agarose gel electrophoretic analysis of one-tube RT-PCR of ApMV-infected and non-infected Turkish hazelnut cultivars. (1–4) Symptomatically ApMV-infected *Corylus* cultivars, (5, 6) non-infected *Corylus* cultivars, (7) negative control, (8) 100 bp ladder molecular size marker, (9) Lambda/HindIII Marker

stranded RNA (dsRNA), and DNA extractions from plant material [62]. Additionally, these techniques provide an efficient and rapid tool for large-scale early screening of plant material, especially in virus elimination programs [63].

4 Conclusions

Plant pathogen elimination using cryotherapy techniques is a newly developing method that can be readily tested with different plant species and cultivars for which cryogenic processes are available. Cryotherapy-based procedures could also be easily applied in basic tissue culture laboratories related in pathogen-free plant production, where they could simplify the application of wide numbers of plant materials, result in notable density of pathogen-free plants, and prevent the difficulties associated with the excision of small shoot tips. Such protocols do not require any specific tools and only marginally add to the time and cost of the traditional procedures of shoot tip culture for pathogen elimination. Furthermore, the use of cryogenic procedures based on vitrification and optimized regrowth of shoot tips should reduce the risk of genetic stability of treated plants.

References

1. Soosaar JL, Burch-Smith TM, Dinesh-Kumar SP (2005) Mechanisms of plant resistance to viruses. Nat Rev Microbiol 3:789–798
2. Lazarowitz SD (2006) Plant viruses. In: Fields BN, Knipe DM (eds) Fields virology, 5th edn. Lippincott, Williams & Wilkins Publishers, Philadelphia, PA, pp 641–706
3. Wren JD, Roossinck MJ, Nelson RS, Scheets K, Palmer MW, Melcher U (2006) Plant virus biodiversity and ecology. PLoS Biol 4:80

4. Hogenhout SA, Ammar E, Whitfield AE, Redinbaugh MG (2008) Insect vector interactions with persistently transmitted viruses. Annu Rev Phytopathol 46:327–359
5. Faccioli VC, Marani F (1998) Virus elimination by meristem tip culture and tip micrografting. In: Hadidi A et al (eds) Plant virus disease control. American Phytopathological Society, Saint Paul, MN, pp 346–380
6. Maliogka VI, Skiada FG, Eleftheriou EP, Katis NI (2009) Elimination of a new ampelovirus (GLRaV-Pr) and Grapevine rupestris stem pitting associated virus (GRSPaV) from two Vitis vinifera cultivars combining in vitro thermotherapy with shoot tip culture. Sci Hortic 123:280–282
7. Benson EE (2007) Cryopreservation of shoot-tips and meristems. In: Day JG, Pennington MW (eds) Cryopreservation and freezing-drying protocols, methods in molecular biology. Humana Press, Totowa, NJ, pp 121–132
8. Wang QC, Cuellar WJ, Rajama ML, Hiraka Y, Valkonen JPT (2008) Combined thermotherapy and cryotherapy for virus eradication: relation of virus distribution, subcellular changes, cell survival and viral RNA degradation in shoot tips to efficient production of virus-free plants. Mol Plant Pathol 9:237–250
9. Wang MR, Li BQ, Feng CH, Wang QC (2016) Culture of shoot tips from adventitious shoots can eradicate apple stem pitting virus but fails in apple stem grooving virus. Plant Cell Tissue Organ Cult 125:283–291
10. Hu GJ, Dong Y, Zhang Z, Fan X, Ren F, Zhou J (2015) Virus elimination from in vitro apple by thermotherapy combined with chemotherapy. Plant Cell Tissue Organ Cult 121:435–443
11. Parker WB (2005) Metabolism and antiviral activity of ribavirin. Virus Res 107:165–171
12. Skiada FG, Maliogka VI, Katis VI, Eleftheriou EP (2013) Elimination of Grapevine rupestris stem pitting-associated virus (GRSPaV) from two vitis vinifera cultivars by in vitro chemotherapy. Eur J Plant Pathol 135:407–414
13. Wang QC, Valkonen JPT (2008) Elimination of two synergistically interacting viruses from sweet potato using shoot tip culture and cryotherapy. J Virol Methods 154:135–145
14. Cai BH, Zhang JY, Qu SC, Gao ZH, Quiao YS, Zhang Z, Zhu F (2008) Preliminary study on the elimination of strawberry mild yellow edge virus from in vitro shoot tips of strawberry cv. Meihou by vitrification cryopreservation. J Fruit Sci 25:872–876
15. Wang QC, Valkonen JPT (2009) Improved recovery of cryotherapy-treated shoot tips following thermotherapy of in vitro-grown stock shoots of raspberry (Rubus idaeus L.). CryoLetters 30:171–182
16. Yi JY, Lee GA, Jeong JW, Lee SY, Lee YG (2014) Elimination potato virus Y (PVY) and potato leaf roll virus (PLRV) using cryotherapy of in vitro grown potato shoot tips. Korean J Crop Sci 59:498–504
17. Pathirana R, McLachlan A, Hedderley D, Carra A, Carimi F, Panis B (2015) Removal of leafroll viruses from infected grapevine plants by droplet vitrification. Acta Hortic 1083:491–498
18. Li BQ, Feng CH, Hu LY, Wang MR, Wang QC (2016) Shoot tip culture and cryopreservation for eradication of apple stem pitting virus (ASPV) and apple stem grooving virus (ASGV) from apple rootstocks 'M9' and 'M26'. Ann Appl Biol 168:142–150
19. Waterworth HE, Hadidi A (1998) Economical losses due to plant viruses. In: Hadidi A et al (eds) Plant virus disease control. American Phytopathological Society, Saint Paul, MN, pp 1–13
20. Esquinas-Alca´ zar J (2005) Protecting crop genetic diversity for food security: political, ethical and technical challenges. Nat Rev Genet 6:946–953
21. Kaya E, Yilmaz Gokdogan E (2015) Virus eradication from plants via novel biotechnological processes: one step freezing methods based on vitrification of cryotherapy techniques. Mugla J Sci Technol 1(2):34–40
22. Helliot B, Panis B, Poumay Y, Swenen R, Lepoivre P, Frison E (2002) Cryopreservation for the elimination of cucumber mosaic and banana streak viruses from banana (Musa spp.). Plant Cell Rep 20:1117–1122
23. Ding F, Jin S, Hong NI, Zhong Y, Cao Q, Yi G, Wang G (2008) Vitrification-cryopreservation, an efficient method for eliminating Candidatus Liberobacter asiaticus, the citrus Huanglongbing pathogen, from in vitro adult shoot tips. Plant Cell Rep 27:241–250
24. Wang QC, Mawassi M, Li P, Gafny R, Sela I, Tanne E (2003) Elimination of Grapevine virus A (GVA) by cryopreservation of in vitro-grown shoot tips of Vitis vinifera L. Plant Sci 165:321–327
25. Bayati S, Shams-Bakhsh M, Moieni A (2011) Elimination of Grapevine Virus A (GVA) by cryotherapy and electrotherapy. J Agr Sci Technol 13:443–450
26. Wang Q, Liu Y, Xie Y, You M (2006) Cryotherapy of potato shoot tips for efficient elimination of Potato leaf roll virus (PLRV) and Potato virus Y (PVY). Potato Res 49:119–129

27. Brison M, de Boucaud MT, Pierronnet A, Dosba F (1997) Effect of cryopreservation on the sanitary state of a cv. Prunus rootstock experimentally contaminated with Plum Pox Potyvirus. Plant Sci 123:189–196
28. Shin JH, Kang DK, Sohn JK (2013) Production of yam mosaic virus (ymv)-free Dioscorea opposita plants by cryotherapy of shoot-tips. Cryo Letters 34(2):149–157
29. Engelmann F (2004) Plant cryopreservation: progress and prospects. In Vitro Cell Dev Biol Plant 40:427–433
30. Mazur P (1984) Freezing of living cells: mechanisms and applications. Am J Physiol Cell Physiol 247:125–142
31. Meryman HT, William RT, Douglas MSTJ (1977) Freezing injury 2solution effects and its prevention by natural or artificial cryopreservation. Cryobiology 14:287–302
32. Engelmann F (1997) In vitro conservation methods. In: Ford-Lloyd BV, Newburry JH, Callow JA (eds) Biotechnology and plant genetic resources: conservation and use. CABI, Wallingford, pp 119–162
33. Ozudogru EA, Capuana M, Kaya E, Panis B, Lambardi M (2010) Cryopreservation of Fraxinus excelsior L. embryogenic callus by one-step freezing and slow cooling techniques. CryoLetters 31(1):63–75
34. Withers LA, Engelmann F (1998) In vitro conservation of plant genetic resources. In: Altman A (ed) Biotechnology in agriculture. Marcel Dekker Inc., New York, pp 57–88
35. Fahy GM, MacFarlane DR, Angell CA, Meryman HT (1984) Vitrification as an approach to cryopreservation. Cryobiology 21:407–426
36. Burke MJ (1986) The glass state and survival of anhydrous biological systems. In: Leopold AC (ed) Membrane, metabolism and dry organisms. Cornell University Press, Ithaca, NY, pp 358–364
37. Fahy GM, Levy DI, Ali SE (1987) Some emerging principles underlying the physical properties, biological actions and utility of vitrification solutions. Cryobiology 24:196–213
38. Kaya E, Alves A, Rodrigues L, Jenderek M, Hernandez-Ellis M, Ozudogru A, Ellis D (2013) Cryopreservation of eucalyptus genetic resources. CryoLetters 34(6):608–618
39. Ozudogru EA, Kaya E (2012) Cryopreservation of Thymus cariensis and T. vulgaris shoot tips: comparison of three vitrification-based methods. CryoLetters 33(5):363–375
40. Ozudogru EA, Kaya E, Kirdok LM (2011a) Comparison of different PVS2-based procedures for cryopreservation of Thymus spp. European Cooperation in Science and Technology, Food and Agriculture, Cryoplanet, COST Action 871:86–92
41. Ozudogru EA, Kirdok E, Kaya E, Capuana M, Benelli C, Engelmann F (2011b) Cryopreservation of redwood (Sequoia sempervirens (D. Don.) Endl.) in vitro buds using vitrification-based techniques. CryoLetters 32 (2):99–110
42. Sakai A, Kobayashi S, Oiyama I (1990) Cryopreservation of nucellar cells of navel orange (Citrus sinensis Osb. var. brasiliensis Tanaka) by vitrification. Plant Cell Rep 9:30–33
43. Steponkus PL, Langis R, Fujikawa S (1992) In: Steponkus PL (ed) Advances in low temperature biology, vol 1. JAI Press, Hamptomill, UK, pp 1–16
44. Sakai A, Kobayashi S, Oiyama I (1991) Cryopreservation of nucellar cells of navel orange (Citrus sinensis Osb.) by a simple freezing method. Plant Sci 74:243–248
45. Nishizawa S, Sakai A, Amano Y, Matuzawa T (1993) Cryopreservation of Asparagus (Asparagus officinalis L.) embryogenic suspension cells and subsequent plant regeneration by vitrification method. Plant Sci 91:67–73
46. Matsumoto T, Sakai A, Yamada K (1994) Cryopreservation of in vitro-grown apical meristems of wasabi (Wasabia japonica) by vitrification and subsequent high plant regeneration. Plant Cell Rep 13:442–446
47. Reinhoud PJ (1996) Cryopreservation of tobacco suspension cells by vitrification. Doctoral Paper, Rijks University, Leiden, The Netherlands
48. Engelmann F (2000) Importance of cryopreservation for the conservation of plant genetic resources. In: Engelmann F, Takagi H (eds) Cryopreservation of tropical plant germplasm—current research progress and applications. JIRCAS/IPGRI, Tsukuba/Rome, pp 8–20
49. Kaya E (2016) Long-term germplasm conservation of two economical important Musa species via cryopreservation-dehydration technıque. Biological Diversity and Conservation 9(3):178–182
50. Kaya E, Souza FVD, Yılmaz Gökdoğan E, Ceylan M, Jenderek M (2016) Cryopreservation of citrus seed via dehydration followed by immersion in liquid nitrogen. Turk J Biol 41:242–248
51. Berjak P, Farrant JM, Mycock DJ, Pammenter NW (1989) Homoiohydrous (recalcitrant) seeds: the enigma of their desiccation sensitivity and the state of water in axes of Landolphia kirkii Dyer. Planta 186:249–261

52. Matsumoto T, Sakai A, Takahashi C, Yamada K (1995) Cryopreservation of in vitro-grown apical meristems of wasabi (Wasabia japonica) by encapsulation vitrification method. CryoLetters 16:189–196
53. Dereuddre J, Hassen M, Blandin S, Kaminski M (1991) Resistance of alginate-coated somatic embryos of carrot (Daucus carota L.) to desiccation and freezing in liquid nitrogen: 2. Thermal analysis. CryoLetters 12:135–148
54. Engelmann F, Takagi H (2000) Cryopreservation of tropical plant germplasm-current research progress and applications. JIRCAS/IPGRI, Tsukuba/Rome
55. Schäfer-Menuhr A, Schumacher HM, Mix-Wagner G (1997) Long-term storage of old potato varieties by cryopreservation of shoot-tips in liquid nitrogen. Plant Genet Resour Newsletter 111(1):19–24
56. Panis B, Piette B, Swennen R (2005) Droplet vitrification of apical meristems: a cryopreservation protocol applicable to all Musaceae. Plant Sci 168(1):45–55
57. Souza FVD, Kaya E, Vieira J, Hilo de Souza E, Amorim O, Skogerboe D, Matsumoto T, Alves C, Ledo S, Jenderek M (2015) Droplet-vitrification and morphohistological studies of cryopreserved shoot tips of cultivated and wild pineapple genotypes. Plant Cell Tissue Org Cult 124(2):351–360
58. Kaya E, Souza FVD (2017) Comparison of two PVS2-based procedures for cryopreservation of commercial sugarcane (Saccharum spp.) germplasm and confirmation of genetic stability after cryopreservation using ISSR markers. In Vitro Cell Develop Biol Plant 53(4):410–417
59. Schaad NW, Frederick RD, Shaw J, Schneider WL, Hickson R, Petrillo MD, Luster DG (2003) Advances in molecular-based diagnostics in meeting crop biosecurity and phytosanitary issues. Annu Rev Phytopathol 41:305–324
60. Van Weemen BK, Schuurs AHWM (1971) Immunoassay using antigen-enzyme conjugates. FEBS Lett 15:232–236
61. Ward E, Foster SJ, Fraaije BA, McCartney HA (2004) Plant pathogen diagnostics: immunological and nucleic acid-based approaches. Ann Appl Biol 145:1–16
62. White EJ, Venter M, Hiten NF, Burger JT (2008) Modified Cetyltrimethylammonium bromide method improves robustness and versatility: the benchmark for plant RNA extraction. Biotechnol J 3(11):1424–1428
63. Kaya E (2015) Using reverse transcription-polymerase chain reaction (RT PCR) for determination of apple mosaic ilarvirus (ApMV) in Hazelnut (Corylus avellana L.) Cultivars. JSM Biochem Mol Biol 3(1):1011

INDEX

Muhammad Naseem and Thomas Dandekar (eds.), *Plant Stem Cells: Methods and Protocols*, Methods in Molecular Biology, vol. 2094, https://doi.org/10.1007/978-1-0716-0183-9, © Springer Science+Business Media, LLC, part of Springer Nature 2020

MIX
Papier aus verantwortungsvollen Quellen
Paper from responsible sources
FSC® C105338

If you have any concerns about our products,
you can contact us on
ProductSafety@springernature.com

In case Publisher is established outside the EU,
the EU authorized representative is:
Springer Nature Customer Service Center GmbH
Europaplatz 3, 69115 Heidelberg, Germany

Printed by Libri Plureos GmbH
in Hamburg, Germany